A Climate
Modelling Primer

RESEARCH AND DEVELOPMENTS IN CLIMATE AND CLIMATOLOGY

Series editors

A. Henderson-Sellers and M. M. Verstraete

A Climate Modelling Primer

A. Henderson-Sellers and K. McGuffie

A Climate Modelling Primer

A. Henderson-Sellers and K. McGuffie

Department of Geography
University of Liverpool

JOHN WILEY & SONS

Chichester · New York · Brisbane · Toronto · Singapore

Library of Congress Cataloging-in-Publication Data:
Henderson-Sellers, A.
A climate modelling primer.

(Research and developments in climate and climatology)
Includes bibliographies and index.
1. Climatology—Mathematical models. I. McGuffie, K.
II. Title. III. Series.
QC981.H5 1987 551.6′0724 86-32505

ISBN 0 471 91462 2

British Library Cataloguing in Publication Data:
Henderson-Sellers, Ann
A climate modelling primer.—
(Research and developments in climate and climatology)
1. Climatology—Simulation methods
I. Title II. McGuffie, K. III. Series
551.6′0724 QC874.5

ISBN 0 471 91462 2

Phototypeset by Dobbie Typesetting Service, Plymouth, Devon
Printed and bound in Great Britain by St. Edmundsbury Press, Bury St. Edmunds, Suffolk

This book is dedicated to all those who believe that daisies grow on Planet McClatchey

Contents

Preface

A primer serves as the *first* means of instruction or 'a prayer-book or devotional manual for the use of the laity' (Oxford English Dictionary, 1973). We set out to write a book to serve as a primer for undergraduates and new graduate students on the basics of climate modelling. Our main motivation is the lack of any single text that can serve as a general introduction to the climate-modelling literature to anyone not previously acquainted with it. We have assumed our readers possess sound high-school skills in mathematics, but in all cases the text can be read without the reader needing to follow the mathematical developments. Throughout we have tried to underline the importance of simple models of the climate. Without these it is impossible to obtain a grasp of the relative importance of different forcing effects. These simple models are also invaluable in testing and extending the concepts upon which much more complex models are founded.

Ten years ago climate modelling was dominated by atmospheric physicists and no one without a sound training in fluid dynamics, radiative transfer or numerical analysis could hope or expect to make a contribution. This scene is changing rapidly. The 'other' subsystems of the climate are now recognized as being at least as important as the atmosphere. Modelling groups are actively searching out oceanographers, ecologists, geographers, glaciologists and remote sensers. To be useful to the climate-modelling community such scientists must acquaint themselves with the fundamental concepts and sensitivities of climate models. There is another group of people for whom an understanding of the basis of climate modelling is becoming important. This group comprises economists, planners, sociologists, demographers and even politicians, who need to know how much credibility should be given to different model types and when and how to apply the results of modelling exercises. *It is for all these people that this book is intended.* We have included a list of recommended reading at the end of each chapter. These reading lists are intended to provide a gateway to the large body of climate-modelling literature. A more complete bibliography is given at the end of the book.

All the programs described are available on IBM PC discs. Details from the authors.

We are only too keenly aware of the generalizations and simplifications that we have made in our attempts to explain the basic mechanisms of climate modelling. We recognize that most climate modellers will take exception to many specific passages. We can only beg their indulgence and point out once more that this book was not really aimed at them. It is perhaps more important for us to illustrate to our intended readership the considerable success achievable with current climate models. This is important because much of this book explains and even emphasizes the incomplete treatment of processes in climate models. In order that our intended readers should not be unduly depressed, we have taken the unconventional step of including a figure in the Preface which compares the predictions made by a range of economic/social models and those made by a range of climate models. For this illustration we have picked two prediction exercises associated with the well-known fossil fuel input of carbon dioxide to the atmosphere and the associated induced temperature rise caused by the greenhouse effect. Figure P.1(a) shows a range of projections of carbon emissions for the year 2050. The present day value for carbon emission is 5 Gt yr^{-1} and the range of predictions for 2050 is 2.5 to more than 25 Gt yr^{-1}. Thus the economic and energy usage models employed in constructing this figure offer a range of predictions from -50% to $>500\%$ of the present day value. By comparison Figure P.1(b) illustrates the predictions of a range of climate models of the CO_2-induced warming caused by a doubling of the atmospheric carbon dioxide level. This hypothesized doubling is in line with the emission figures shown in Figure P.1(a) inasmuch as the predicted emissions for 2050 would give rise to approximately 370 to 530 ppmv of atmospheric carbon dioxide, which is between 1.4 and 2.0 times the pre-industrial level of approximately 270 ppmv. The predictions of additional temperature increase by the climate models range from about 2 to 7 °C. These predictions can also be expressed as percentages of the present day greenhouse increment which is 33 °C. The climate modellers' predicted range is between $+6\%$ and $+21\%$ which is a much closer agreement than is offered by the energy use/economic models. The most recent evaluation of the temperature increase likely to be caused by doubling of atmospheric CO_2 (World Meteorological Organization (WMO), 1986) is a temperature increase of between 1.5 and 4.5 °C so that it can be seen that the differences amongst climate model predictions is narrowing rapidly. We hope that our readers will bear this agreement in mind when considering the specific examples of climate model results given here and those in the considerable volume of climate modelling literature. The Earth's climate is a complex and highly non-linear system. The important result of this is that as models increase in complexity, individual numerical experiments can produce unexpectedly disappointing simulations often because only part of a complex process has been captured.

Climate modelling can be great fun. It is so broad and so interdisciplinary that you are forced to acquire new skills and understand new techniques. We recommend it as a new approach to many problems, as a pastime and perhaps even as a career.

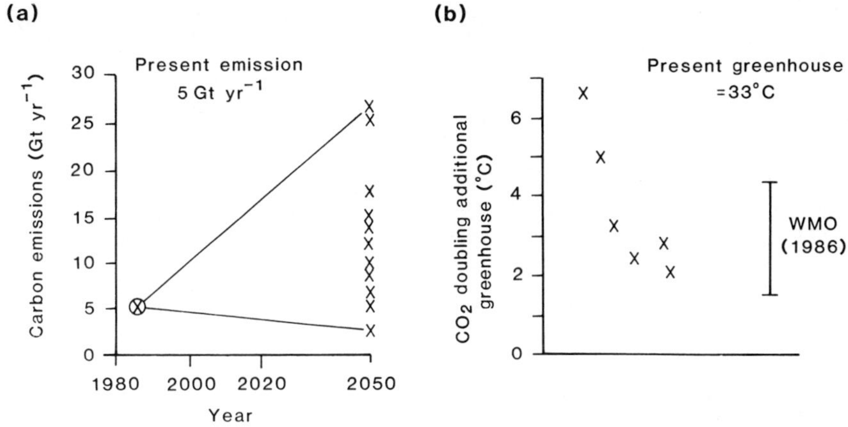

Figure P.1. (a) Predictions of the emission rates of carbon for the year 2050 made with a range of energy/economic models (described by Keepin *et al.*, 1986). The range of estimates is from −50% to +500% of the present day emission. (b) Predictions of the additional greenhouse increment caused by doubling the level of atmospheric carbon dioxide. The model types (described by Spelman and Manabe, 1984) include both restricted and global domains and idealized and realistic geography. The recent 'consensus' prediction (WMO, 1986) is also shown. The range of estimates is from +6% to +21% of the present day greenhouse effect of 33 °C.

A. Henderson-Sellers and K. McGuffie,
Liverpool, October 1986.

Acknowledgements

We are exceedingly grateful to all our friends and colleagues who read through parts or all of various drafts of this book. In particular we want to thank Dave Carson, Bob Dickinson, Graham Cogley, Graham Thomas, Mo Wilson, Peter Briggs and Mary Benbow and all the 'class of 1986' in Liverpool.

We reckon we ought to thank each other—she should thank him, especially, for not allowing a bad bout of food poisoning to interfere with proofing and he should thank her for everything he had to relearn about electronics when, at a crucial stage, she shorted out the computer keyboard by tipping her screwdriver over it.

Brian, as usual, deserves the biggest 'thank you' of all for helping, advising and contributing more positively than we deserved. Finally, we want to thank Mynhed for suffering, apparently gladly, 'grousin' grids' and 'numerical screams'.

Chapter 1

Climate

'The climate is a beautiful system, exceedingly rich in interconnections and complexities.'
A. H. Oort (public communication, 1986)

1.1 THE COMPONENTS OF CLIMATE

The term 'climate' has a very wide variety of meanings. To a geologist or geomorphologist the 'climate' is an external agent which forces many of the phenomena of interest. For an archaeologist the 'climate' of an earlier time might have been a crucial influence upon the people being studied, or might have been of little socio-economic significance but still so strong an environmental feature that it has left a 'signature' which can be interpreted. An agriculturalist probably sees the 'climate' as the background 'norm' upon which year-to-year and day-to-day weather is imposed, while the man in the street may speak of moving to a location with a 'better climate'. To many of us, 'climate' often first suggests temperature although rainfall and humidity may also come to mind. When we think of climatic change it is often in the time frame of glacial periods. More recently, however, there has been considerable public concern over the possible shorter-term impact upon the climate of increasing atmospheric carbon dioxide and other trace greenhouse gases.

The climate is both a forcing agent and a feature liable to be disturbed. It can fluctuate on relatively short timescales, producing for example the Sahel and Ethiopian droughts, and over much longer times giving rise to glacial epochs. The climate is perceived in terms of the features of the entire climate system which most readily or most usefully characterize the phenomenon of interest. All of these characteristics of the climate are depicted in Figure 1.1. The three axes themselves are fundamental but the intervals are arbitrary and many more could be included.

A single satisfactory definition of climate is probably unobtainable because the climate system encompasses so many variables and so many time- and space-scales. A useful definition might be 'all of the statistics of a climatic state determined over an agreed time interval (seasons, decades or longer), computed for the globe or possibly for a selected region'. This definition is not especially helpful because it is so broad, although it does serve to emphasize the usefulness

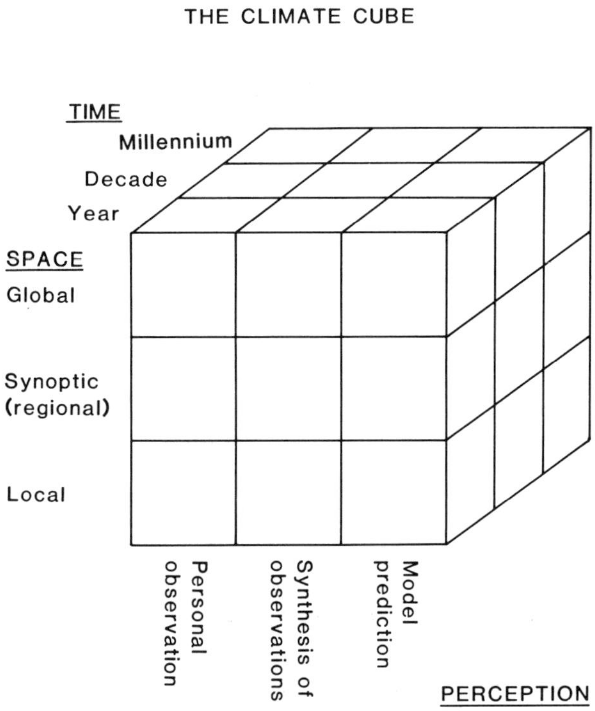

Figure 1.1 The climate cube. Climate can be viewed as existing in at least three domains: time, space and human perception. The divisions of these domains depicted here are arbitrary, a great many more could exist. Historically, individual disciplines have been concerned with single 'cells'. Recently the extent of the climate system and the importance of interactions between domains have been recognized.

of higher-order statistics such as variance that can often be more useful in characterizing a climatic state than the mean. The definition also permits further description of a climatic change as the difference between two climatic states, and a climatic anomaly as the difference between a climatic state and the mean state.

Introduction and outline of the book

In this book we have set out to introduce and describe the way in which the climate is modelled. By modelling we mean the simulation using physically based formulations of the processes which make up the climate system. We are concerned with explaining the approaches and methods employed by climate modellers and shall not be concerned directly with meteorology, socio-economic impacts of climatic changes or palaeoclimatic reconstruction, although all of these disciplines and many others will be drawn upon in our descriptions.

In this first chapter we identify the components of the climate system and the nature of their interactions, as well as describing briefly some of the motivations of climate modellers. Chapter 2 contains a history of climate modelling and provides an introduction to all the types of models to be discussed in subsequent chapters. The other chapters are concerned with different model types, their development and applications. Throughout we have taken climate models to be predictive descriptions of regional- to global-scale phenomena; hence empirically based 'models' such as crop prediction equations and water resource management codes have not been included. The reason for this limitation is not that such models are uninteresting but rather that they have grown from well-identified fields and thus background literature can be readily obtained. Climate modelling in the sense in which we use the term, on the other hand, has developed from a wide variety of backgrounds in a somewhat haphazard manner and consequently there is little background to which the uninitiated can refer.

In one sense the book develops the background material required for understanding of the most complex type of climate model, the general circulation climate model, by illustrating principles in other, simpler, model types. Thus it is necessary to introduce the concept of energy balance, especially planetary radiation balance, before one-dimensional energy balance models (Chapter 3) can be understood. The other one-dimensional model type, the radiative–convective model (Chapter 4), is based upon the transfer of both shortwave (solar) radiation and longwave (terrestrial) radiation through the atmosphere. In this model type the concept of vertical motion forced by heating from below (i.e. convection) is required. In Chapter 5 the importance of the large-scale meridional flow of the atmosphere is underlined by demonstrating that two-dimensional statistical dynamical models, which are formulated in terms of this flow, are, in many senses, comparable with general circulation climate models. Hence by Chapter 6 the reader is, hopefully, well prepared to understand the way in which radiative forcing, vertical motion and horizontal advection are included in three-dimensional models of the climate.

On the other hand, most of this preparation has involved incorporation of atmospheric processes. To understand the whole of the climate system it is necessary to include sub-models of the other components too. Throughout the book an effort will be made to underline the importance of simpler models in understanding the complex interactions between various components of the climate system. Complex models are only one, particularly sophisticated, method of studying climate. They are not necessarily the best tool and simple models are often used in conjunction with, or sometimes even to the exclusion of, more complex and apparently complete models. The literature is full of examples of very simple models being used to demonstrate failures in much more complex systems.

Last, but by no means least, any introduction to climate modelling must stress the crucial role played by computers. Without the recent growth in computational power and the reduction in computing costs most of the developments in climate modelling which have taken place over the last two decades could not have happened. We have intentionally emphasized computing tools over mathematical skills in the description of the simplest type of climate model, the energy balance model (EBM), in Chapter 3. In this chapter the full BASIC code of an EBM is listed and time is taken to develop the type of questions that might prompt further use and thought. A similar, but more complex, code could be developed for a one-dimensional radiative–convective (RC) model. It is estimated that a full atmospheric general circulation model (GCM) takes about 25–30 person-years to code up and the code requires continual updating as new ideas are implemented and as advances in computer science are accommodated.

We have included two appendices to the main text. Appendix A contains a glossary of terms which may be new to readers unfamiliar with climatology/meteorology. As we have used this glossary for definitions rather than disturb the main thread of the text, reference to it is recommended. Appendix B contains a description of the implementation and use of four simple climate models. The model codes (which were designed for an IBM PC) can be purchased on a floppy disc. Full details are given in the back of the book. The four climate models have been chosen to illustrate fundamental concepts of the climate system: (i) feedback, specifically biospheric feedback in the example code; (ii) the impact of changing one feature of the climate system, here albedo change is considered; (iii) radiation and clouds examined through the use of an energy balance climate model; and (iv) the difficulties and possible results of coupling ocean characteristics into a climate model; in the selected code the model is a simple box model.

The climate system

The climate system was defined, in a document produced by the Global Atmospheric Research Programme (GARP) of the World Meteorological Organization in 1975, as being composed of the *atmosphere, hydrosphere, cryosphere, land surface and biosphere*. In the same document the representation of the climate shown in Figure 1.2 was given. This diagram complements Figure 1.1 by emphasizing the components and processes rather than the space- and timescales. The diagram and the order of the climatic components listed (and the name of the organization which produced the list) underline an important feature of climate modelling: its early phases were driven by atmospheric scientists. The order of the components of the climate enumerated in 1975 is also a rough indicator of the order in which these elements were considered and, to some extent, their relative time-scales. The first modelled component was the atmosphere, which, because of its low density and ease of movement, is the most 'nervous' of the climatic subsystems. Precipitation has been included but clouds are difficult to incorporate successfully, and linking the major part of the hydrosphere, the oceans, into climate models is proving very difficult. This is partly because the critical space- and timescales of the ocean and atmosphere subsystems differ but also because the coupling between the subsystems is strongly latitude-dependent. In the tropics the systems are closely coupled especially through temperature (Figure 1.3), in mid-latitudes the coupling is weak, predominantly

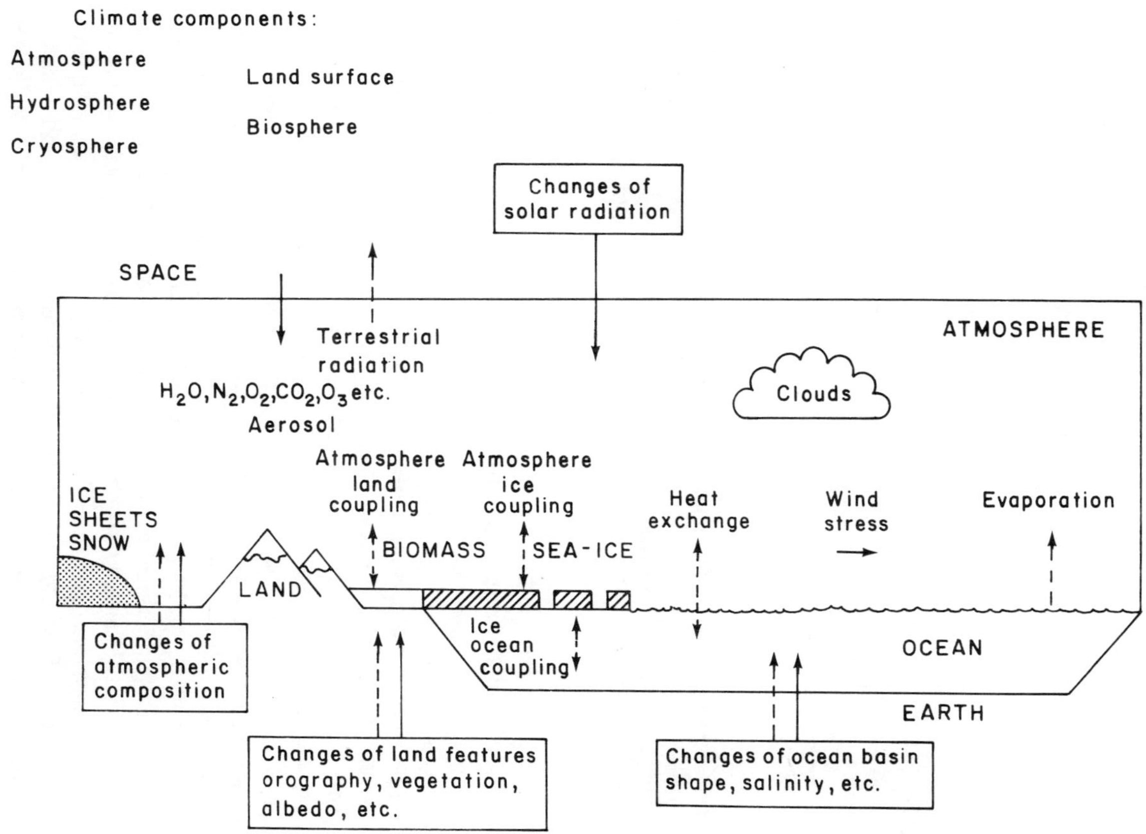

Figure 1.2 Schematic illustration of the components of the climate system. Full arrows are examples of external processes and dashed arrows are examples of internal processes (reproduced with permission from GARP, 1975).

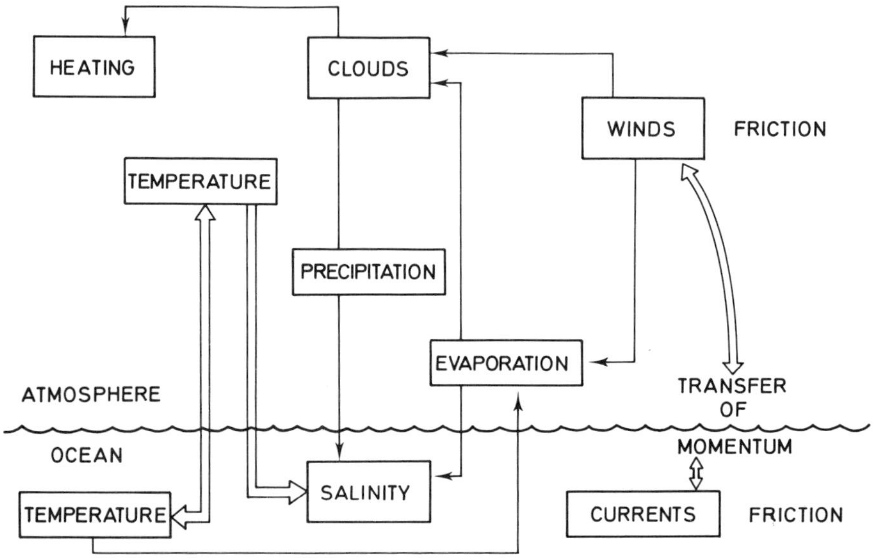

Figure 1.3 A representation of the major coupling mechanisms between the atmosphere and ocean subsystems. The relative importance of these coupling mechanisms varies with latitude. The feedback between atmospheric temperature and oceanic salinity is interesting because it is strong only in the sense of the atmosphere forcing the ocean.

via momentum transfer, but in high latitudes there is a close coupling, primarily through salinity, which is responsible for the formation of oceanic deep bottom water.

The cryosphere has been incorporated into climate models to some extent following the description of simple EBMs, but it has recently been recognized that the insulating effect of the cryosphere is at least as important as its albedo effect. Sea-ice, especially, decouples the ocean from the overlying atmosphere causing considerable changes in both subsystems.

The final two components were originally added almost as an afterthought. Scientists concerned with land-surface processes had described the climate as both an agent and a feature of change for over a century before the new breed of climate modellers began serious consideration of their theories. The importance of the biosphere was underlined by the realization, following the first studies of climate impact, that the carbon dioxide component of the atmosphere is critically dependent upon the biota. These studies are generally directed towards the state of the continental surface and the growth and well-being of plants and animals.

In this rather clumsy fashion and from mixed parentage the science of climate modelling was born. Climate modellers were to discover very quickly that the system that they had summarized so neatly in a figure and a list of components was exceedingly complex, containing links and feedbacks which are highly non-linear and hence difficult to identify and reproduce.

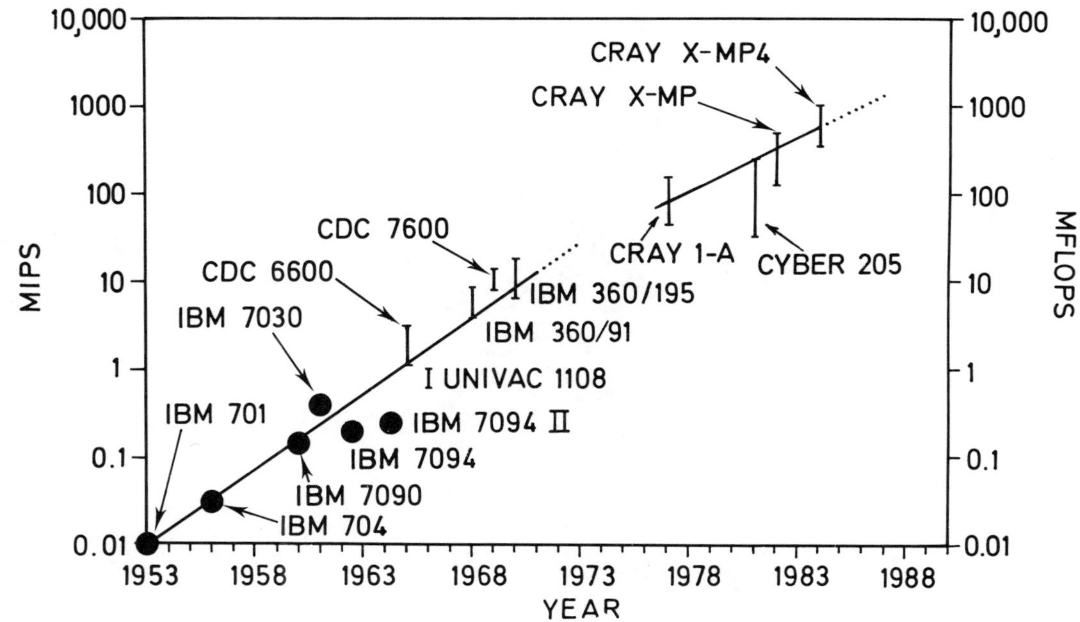

Figure 1.4 Performance of the most powerful computers between 1953 and 1984. The power is given in millions of instructions undertaken per second (MIPS) up to 1975 and, more recently, in millions of floating point operations per second (MFLOPS). (Note the vertical scale is logarithmic.) The increase in power was almost exponential at first and, although improvement has slowed, it is still very impressive (reproduced with permission from Simmons and Bengtsson, 1987).

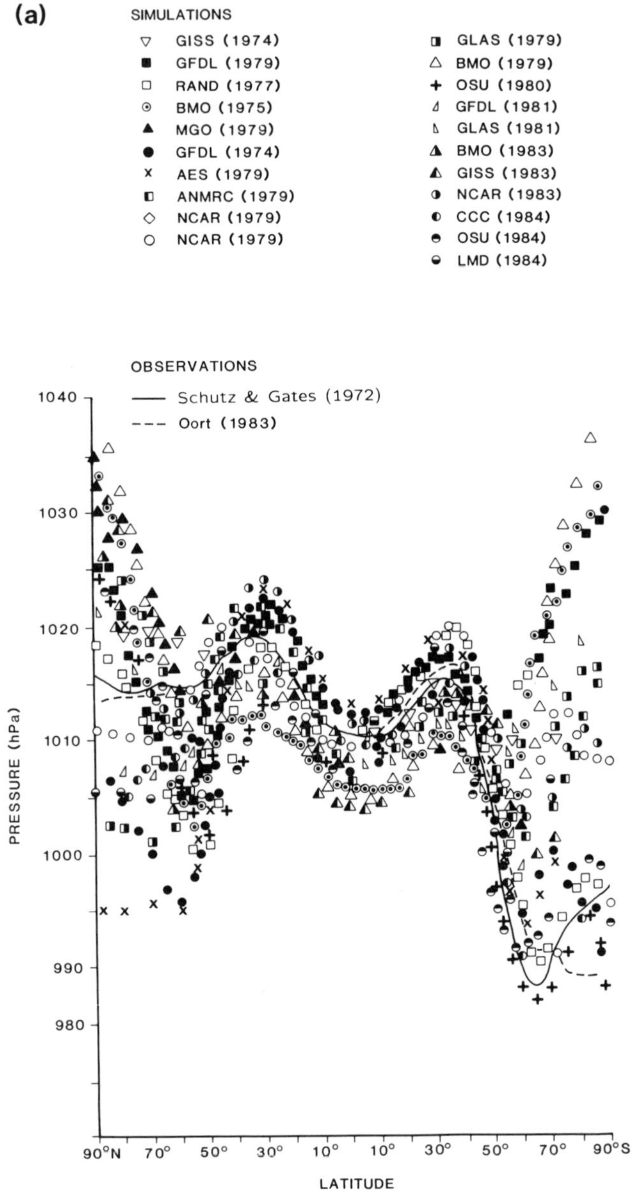

Figure 1.5(a) (*caption opposite*)

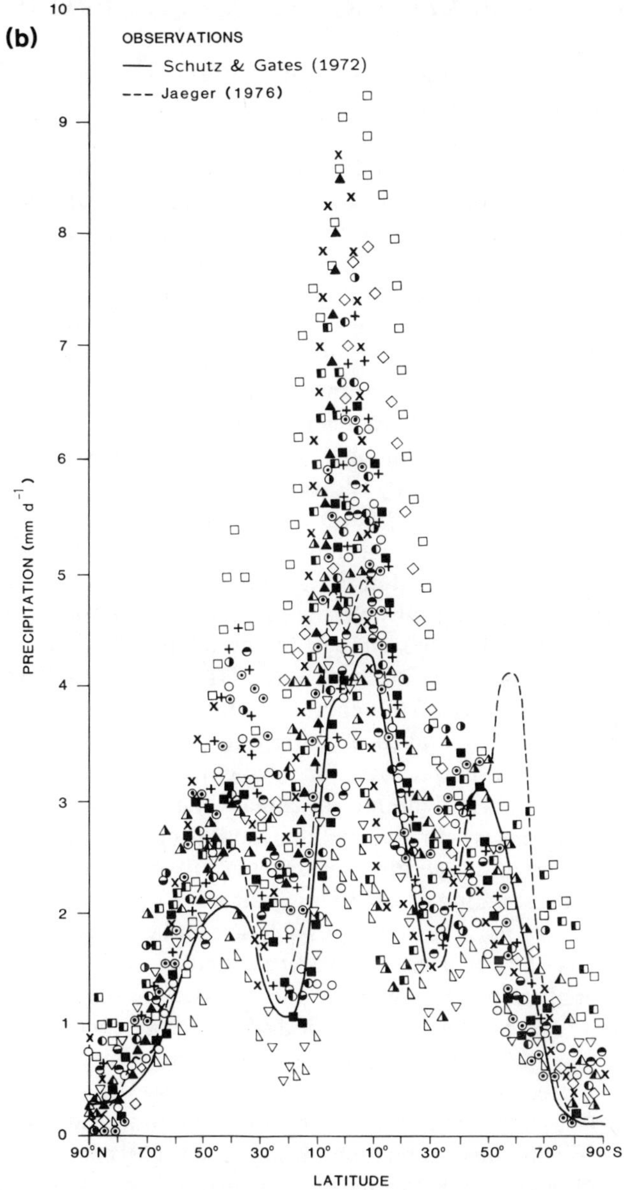

Figure 1.5 The performance of atmospheric GCMs in terms of their simulation of: (a) sea-level pressure; and (b) precipitation. The range between different models is great but has reduced over the years. It is also important to recognize that these (and other) observational data are incomplete and, probably, somewhat inaccurate and that the "improving" trend can be dramatically if temporarily reversed in specific model simulations when only part of a process has been captured (reproduced by permission from Gates, 1985).

1.2 CLIMATE MODELLING AND CLIMATE PREDICTION

From the viewpoint of the climatologist the 'carbon dioxide debate' epitomizes modern climatology. Society asks for predictions of the future climate and for information about the potential impacts of climatic changes. Such requests provide the challenges and opportunities of climatology. It is through a physically based computer-modelling approach that the questions of society must be answered. There have been great advances made in such modelling over the past 30 years, but even the most sophisticated models are far removed in complexity from the real climate. Further advances are possible, but they need to be associated with increased understanding of the nature of interactions within a model and indeed within the real climate system. Perturbations caused by everything from air pollution to volcanoes, from solar luminosity to climatically induced variation in surface character must be considered. Modelling in such a widely ranging subject is a formidable task and it requires co-operation between many disciplines if reliable conclusions are tobe drawn.

Available computing power has also increased greatly over the past 30 years (Figure 1.4). This rapid increase has meant that climate models have expanded both in terms of complexity and in total simulation time. Multi-year, seasonal simulations are not uncommon even with the most complex global climate models. Figure 1.5 shows the change in model performance over the last 10 of these years. The general trend is that the more recent models attain more accurate simulations of the climate system, as far as is represented by observations of zonally averaged pressure and precipitation.

As our knowledge increases, more facets of climatology will be incorporated into climate models. New insight into the climate system so gained can be utilized in controlling activities which adversely influence the environment. Better and more extensive observations permit processes to be incorporated more fully where a full physical understanding or sufficient computing power is lacking. Certainly faster computers help to increase the possible size and complexity of models, but simple models may be sufficient to answer particular, well-specified problems.

Whether its predictions are 'correct' is the ultimate test of any model. Weather forecast models can be tested over a period of a few days, but models of climate are required to predict decades or even centuries in advance. More importantly, climate model "predictions" offer only a general case of the response since the model climate loses its association with the initial conditions within a few weeks. Hence testing in this sense is virtually impossible. The climate models which will be discussed in this book cover a wide range of space- and timescales. These different types of climate models attract interest from many different scientific disciplines. Long-period modelling may attract glaciologists, geologists or geophysicists. Atmospheric chemists may be interested by detailed modelling of the vertical structure of the atmosphere. Implications of solar-system-scale phenomena attract planetary physicists and astronomers. The contributions of these disciplines enhance the science of climate modelling and in this book we will attempt to show how these contributions fit together. The remainder of this chapter is devoted to a description of the characteristics of the climate system which are of interest to modellers.

1.3 CLIMATE CHANGES AND HUMAN PERCEPTION

Climate change, for example in response to increasing atmospheric CO_2 and trace gases, will probably be small and gradual. Detection is very difficult in the early stages if only one response is monitored. Recently a ''fingerprint'' method has been proposed. Using this detection technique a set of small changes in a number of variables would be required before a signal could be established.

Any changes in climate, whatever the cause, must have an impact on human activity. Crop-yield models indicate that food production depends on the weather. It might therefore be postulated, for example, that a change in climate would have led to consistently low yields in a particular area, which in turn would have led to a human response in terms of a change in agricultural practice. Such simple postulates can be very misleading, since they conceal several problems which are inherent in relating climate change to human impact. These concern the nature of climatic changes themselves, the strength of the relationship between climate changes and human response and the availability of past climatic data.

It is possible to think of climatic changes as being represented by changes in the long-term mean values of a particular climatic variable. Superimposed on this changing mean value will be decadal fluctuations and year-to-year variations. Such short-period variations may, of course, be influenced by the change in the mean. On the human timescale, changes in the mean value are likely to be so slow as to be almost imperceptible. For example, the changes over the last few decades can only be detected by careful analysis of instrument records. Much more noticeable will be variability, expressed, for example, as a 'run of bad winters'. Any human response will depend on such a perception, whether conscious or subconscious. A 'large' climate change may not lead to any response whereas a much smaller change in a particular feature, expressed as a perceived change in variability, may have a profound impact on human activity.

Even if a change occurs which potentially has a significant impact on human activity, a societal response will not necessarily follow. Any response to a climate change is governed by a host of non-climatic factors which need to be considered. Consequently a direct link between climate change and human activity is often difficult to establish.

Any attempt to establish the impact of past climatic changes must use historical information. Pre-instrumental historical records are qualitative and selective and emphasize information about unusual conditions which were perceived as having an impact. Consequently they tell us less about normal conditions. A great deal must therefore be inferred about the climate and its variability before any suggestions regarding its impacts are made.

It has been suggested that during this period (*c*. 1450–1850) average temperatures in mid-latitudes were about 1 K below their present levels. This suggests that the growing season was relatively shorter. This might imply that crop yields were low. However, additional factors such as the incidence of late spring frosts, or destructive winds near harvest time, of which we have little or no information, may have been more significant. It has also been suggested that cold, damp winter conditions prevailed in the Little Ice Age, grain storage became impossible and famine susceptibility increased. Population pressure and plague could have been equally important in creating the problems of this period.

Such an effect is seen on a local scale, in the Lammermuir Hills in Scotland (Figure 1.6). The effect of the temperature decline during the course of the Little Ice Age and the recovery following it is clearly illustrated. The combined isopleths of 1050 degree days plus 60 mm potential water surplus (at the end of the summer) which represents the approximate 'cultivation limit' are seen to expand and move downslope during the cooling period (Figure 1.6(a)) and then to be restored towards the end of the Little Ice Age (Figure 1.6(b)). A careful study of the records of farming and settlement in this area gives credence to the suggestion that in marginal regions human response to climatic deterioration is identifiable. In the centre of an agricultural region farming practices will be well adapted to that particular climate and year-to-year variations will pose little threat. As the margins are approached, however, variability will become more significant. Usually overall production will be low, so that little surplus can be stored against the poor years that climate variability will inevitably bring. If a climate change occurs which alters the frequency of the poor years, some human response must follow.

The definition of 'marginality', of course, depends upon the climatic regime and agricultural practices considered. Three such areas using different climatic indices for marginality are shown in Figure 1.7. Northern Europe is divided into agriculturally marginal and sub-marginal areas. The limits are given by a combined index which is a function of the number of months with a mean monthly temperature above 10 °C and the degree of the precipitation deficit over evapotranspiration, if any, in the summer months. The region of marginal cultivation identified for the United States is based upon total rainfall in the period April–September rather than upon the combination of temperature and rainfall used for northern Europe. For Australia three zones of marginality, for different climate regimes and agricultural enterprises, are shown in Figure 1.7(c). Here the limits are based on temperature and precipitation values and their ranges. The changes of these limits with time indicate the eastward encroachment of aridity and the establishment of new marginal areas.

Future human responses to climatic change are likely to involve similar complex chains of decisions. We will examine the nature of the climatic part of these problems. Consideration of the nature of the interaction of elements of the climate system is an important prerequisite to consideration of potential human response. It is the nature of the climate system and the various degrees to which it can be examined which will be addressed in forthcoming sections.

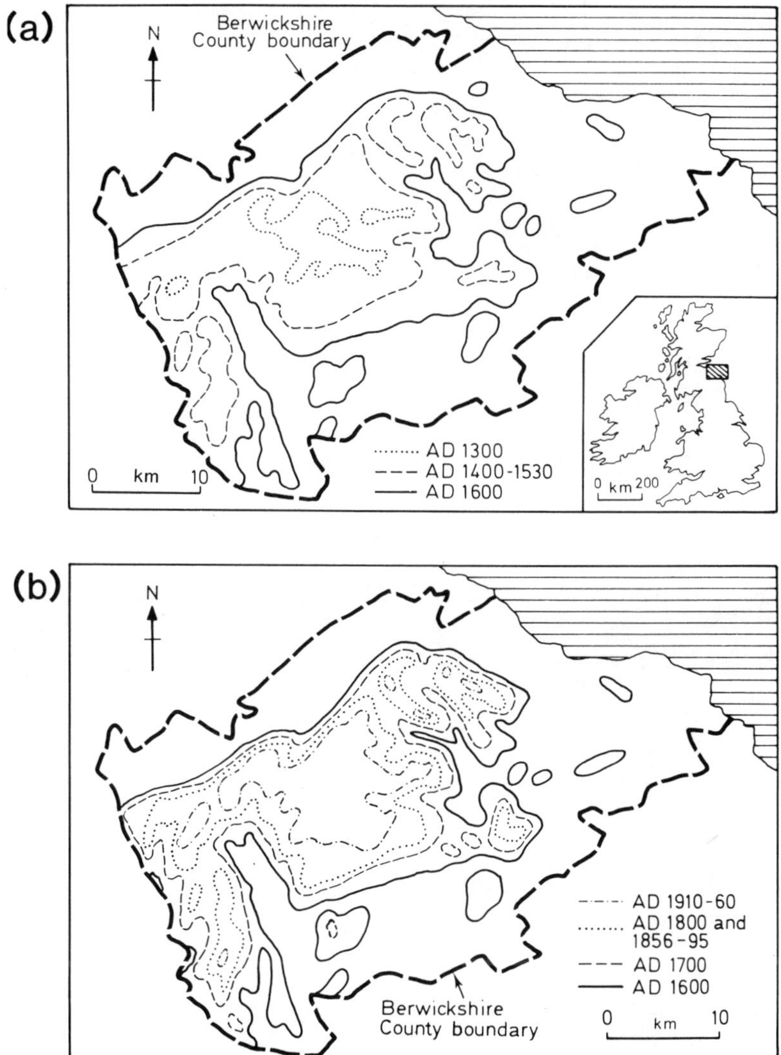

Figure 1.6 (a) Cooling in the period AD 1300 to 1600 causing the 'fall' of the combined isopleths of 1050 degree days and 60 mm potential water surplus in southeast Scotland; (b) warming in the period 1700 to 1910–60 resulting in the 'rise' of the same combined isopleth (reproduced with permission from Parry, 1978).

1.4 FEEDBACK MECHANISMS IN CLIMATE

In the broadest sense a feedback occurs when a portion of the output from the action of a system is added to the input and subsequently alters the output. The result of such a loop system can either be an amplification of the process or a dampening. These feedbacks are labelled positive and negative respectively.

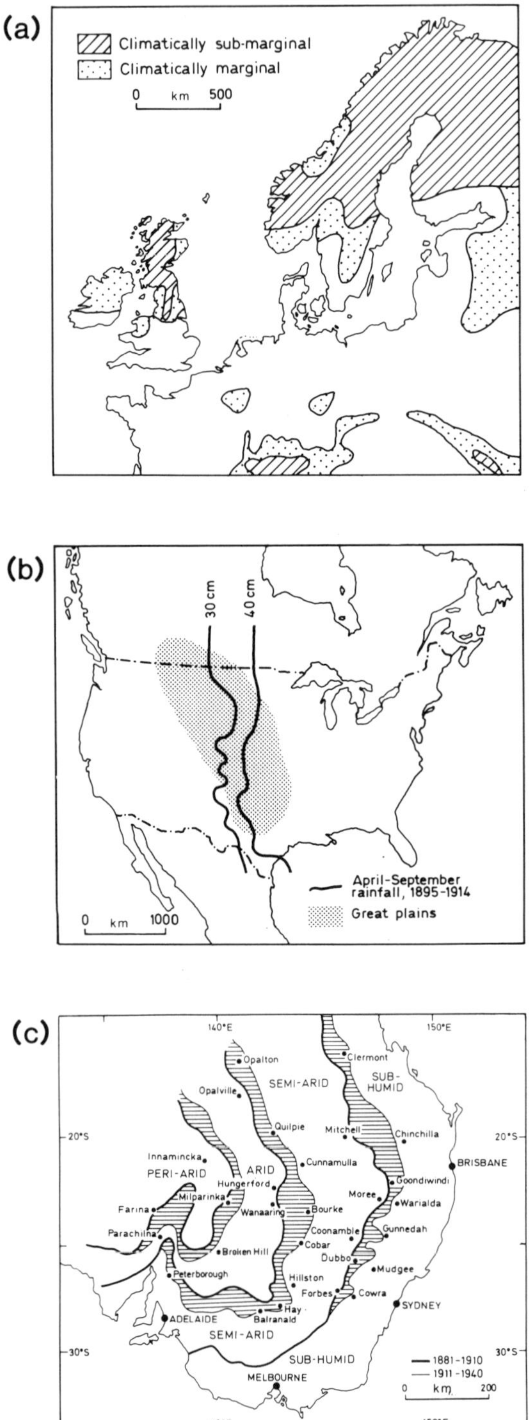

Figure 1.7 Climatologically marginal land in (a) northern Europe; (b) the Great Plains of the USA; and (c) eastern Australia. In (c) the shifts in climatic belts between 1881–1910 and 1911–1940 are seen. (Reproduced with permission from various sources including Gentilli (1971), Elsevier Science Publications).

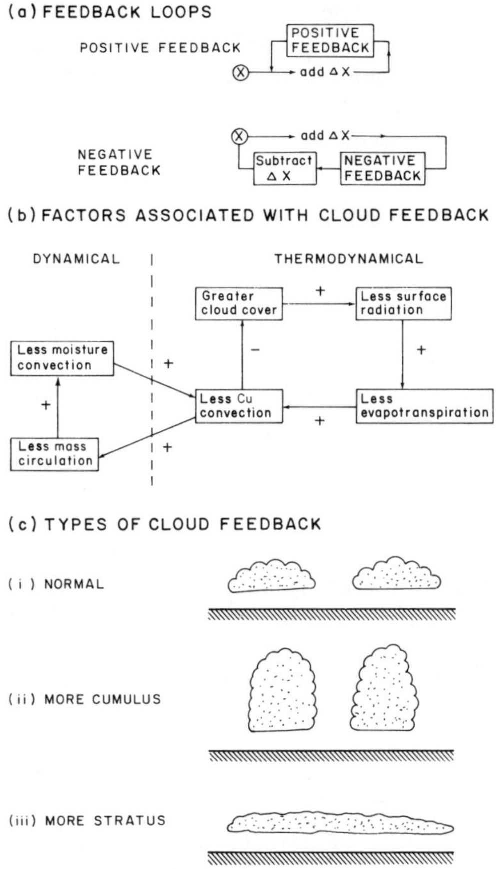

Figure 1.8 (a) Flow diagram illustrating positive and negative feedbacks; (b) specific examples of dynamical and thermodynamical feedbacks and their directions in the case of a change in the amount of cumulus convection; (c) the exact nature of an increase in cloud amount is unclear. The cloud could either be more extensive vertically or more extensive horizontally (reproduced by permission of Longman from Henderson-Sellers and Robinson, *Contemporary Climatology*, 1986).

Positive feedbacks enhance a perturbation whereas negative feedbacks oppose the perturbation (Figure 1.8(a)).

The importance of the direction of a feedback can be simply illustrated by considering the impact of self-image on diet. Someone slightly overweight who eats for consolation can become depressed by the increased calorie intake and so eat more and rapidly become enmeshed in a detrimental positive feedback effect. On the other hand, perception of a different kind can be used to illustrate negative feedback. As a city grows there is a tendency for immigration but the additional influx of industry, cars and people is often detrimental to the environment so that it may be balanced, or even exceeded, by an outflux of richer inhabitants. One of the models in Appendix B illustrates that negative feedback, imposed by the biosphere, can lead to temperatures constrained within a narrow range despite the imposition of external perturbations. In this section some of the feedback mechanisms inherent in the climate system are described. This precedes a discussion of some of the mechanisms which can impose a perturbation upon this system.

1.4.1 The ice-albedo feedback mechanism

Given that some external or internal perturbation acts to decrease the global surface temperature then the formation of additional areas of snow and ice is likely. These cryospheric elements are bright and white, reflecting almost all the solar radiation incident upon them. Their albedo (ratio of reflected to incident radiation) is therefore high. The surface albedo, and probably the planetary albedo (the reflectivity of the whole atmosphere plus surface system as seen from 'outside' the planet), increases. Thus a greater amount of solar radiation is reflected away from the planet and temperatures decrease further. A further increase in snow and ice results from this decreased temperature and the process continues. This positive feedback mechanism is known as the *ice-albedo feedback mechanism*. Of course, this ice-albedo feedback mechanism is also positive if the initial perturbation causes an increase in global surface temperatures. With higher temperatures, the areas of snow and ice are likely to be reduced, so reducing the albedo and leading to further enhancement of temperatures.

1.4.2 The water vapour "greenhouse"

Another positive feedback mechanism occurs with the increase of atmospheric water vapour resulting from increased evaporation as temperatures are raised. Many atmospheric gases contribute to the greenhouse warming of the surface as a result of their absorption of infrared radiation emitted from the surface. The dominant greenhouse gas in the Earth's atmosphere is water vapour, although carbon dioxide and other trace gases, such as chlorofluorocarbons, are becoming increasingly important. The additional greenhouse effect of the extra water vapour enhances the temperature increase. Similarly as temperatures fall (possibly as a result of the ice-albedo feedback) the atmosphere is able to hold less water vapour and the greenhouse effect is reduced.

1.4.3 Cloud feedbacks

To establish even the direction of the feedback associated with clouds is difficult, since they are both highly reflective (thus contributing to the albedo) and composed of water and water vapour (thus contributing to the greenhouse effect). It has been suggested that for low- and middle-level clouds the albedo effect will dominate over the greenhouse effect, so that increased cloudiness will result in an overall cooling. On the other hand, cirrus clouds which are fairly transparent at visible wavelengths have a smaller impact upon the albedo so that their overall effect is to warm the system by enhancing the greenhouse effect.

Cloud feedback is, however, not this straightforward. There are dynamical and thermodynamical factors to be considered (Figure 1.8(b)) so that it is uncertain whether an increased temperature will lead to increased or decreased cloud cover (as opposed to cloud amounts). Although it is generally agreed that increased temperatures will cause higher rates of evaporation and hence make more water vapour available for cloud formation, the form which these additional clouds will take is much less certain. For the same 'volume' of new cloud, an increased dominance of cumuliform clouds probably reduces the percentage of the surface covered by clouds. Stratiform clouds, on the other hand, would increase the area covered (Figure 1.8(c)). Thus using the simplest reasoning it might be claimed that an increase in cumuliform clouds implies positive feedback, whereas an increase in stratiform clouds implies a negative feedback.

The second unknown factor about how clouds change in response to a climate perturbation is their height of formation. The positive feedback observed in some model simulations for a given perturbation is greater if clouds of different types are assumed to occur at fixed heights in the atmosphere. On the other hand it is probably more reasonable to argue that, as cloud formation processes are a strong function of temperature, cloud parameterization schemes should be based upon an assumption of cloud formation taking place at fixed temperatures rather than at fixed heights.

1.4.4 Combining feedback effects

The power of individual feedback effects can be very great, as is illustrated in the fourth model described in Appendix B. Also, and perhaps more importantly, more than one feedback effect is likely to operate within the climate system in response to any given perturbation and these feedback effects often combine in a complex way. For example, consider a system in which a change of surface temperature of magnitude ΔT is introduced. Given no internal feedbacks then this temperature increment will represent the change in the temperature. If feedbacks occur then there will be an additional temperature change and the new value of the temperature change will be

$$\Delta T_{\text{final}} = \Delta T + \Delta T_{\text{feedbacks}} \tag{1.1}$$

where $\Delta T_{\text{feedbacks}}$ can be either positive or negative.

The value of ΔT_{final} (i.e. whether it is large or small) is usually related to the perturbation which caused it by a measure of the sensitivity of the climate system to disturbance.

There are a number of such sensitivity parameters in the climate-modelling literature. The earliest measure of a model's sensitivity was the β parameter which is equal to the ratio of the calculated surface temperature change to an incremental change in the prescribed incident solar radiation. More recently, equation (1.1) has been rewritten in terms of a feedback factor, f, so that

$$\Delta T_{\text{final}} = f\,\Delta T \tag{1.2}$$

This feedback factor has, in turn, been related to the amplification or gain, g, of the system which is defined, using the analogy of gain in an electronic system, by

$$f = 1/(1 - g) \tag{1.3}$$

The 'f' factor is neither additive nor multiplicative, and is thus not an especially useful parameter. Gain factors, g, are additive but depend, as does the β parameter, on knowing the present climate system albedo or outgoing fluxes. A much more convenient climate sensitivity parameter is given in terms of a perturbation in the global surface temperature ΔT which occurs in response to an externally prescribed change in the net radiative flux crossing the tropopause, ΔQ

$$C[\delta(\Delta T)/\delta t] + \lambda\,\Delta T = \Delta Q \tag{1.4}$$

Here $\lambda\,\Delta T$ is the net radiation change at the tropopause resulting from the internal dynamics of the climate system, t is time and C represents the system heat capacity.

Although equation (1.4) represents an extreme simplification of the system it is useful in interpreting and summarizing the sensitivity of more complex climate models. A convenient reference value for λ is the value λ_B which λ would have if the Earth were a simple black body with its present-day albedo,

$$\lambda_B = 4\sigma\,T_e^3 = 3.75 \text{ W m}^{-2}\text{ K}^{-1} \tag{1.5}$$

where σ is the Stefan-Boltzmann constant and T_e is the Earth's effective temperature, both of which are defined in the glossary and explained in Chapter 3. The overall climate system sensitivity parameter λ_{TOTAL} is composed of the summation of λ_B and all contributing feedback factors λ_i such that, for example,

$$\lambda_{\text{TOTAL}} = \lambda_B + \lambda_{\text{water vapour}} + \lambda_{\text{ice albedo}} \tag{1.6}$$

Thus for a given system heat capacity a positive value of feedback factors (λ_i) implies stability or negative feedback and a negative value implies positive feedback and possibly growing instability.

To establish the resulting temperature change the inverse of the value of λ_{TOTAL} is multiplied by the value of the ΔQ for the perturbation considered. The relationship (derived directly from equation (1.4) for the case of zero temperature change) that $\Delta T = \Delta Q/\lambda$ has given rise to another definition of a feedback factor as $1/\lambda$ or λ'. For doubling atmospheric CO_2 it has been shown that $\Delta Q \simeq 4.2$ W m^{-2}. If we take $\lambda_B = 3.75$ W m^{-2} K^{-1} and $\lambda_{water\ vapour} = -1.7$ W m^{-2} K^{-1} and $\lambda_{ice\ albedo} = -0.6$ W m^{-2} K^{-1} we have that $\lambda_{TOTAL} = 1.45$ W m^{-2} K^{-1} so that the globally averaged temperature rise due to doubling atmospheric CO_2 is found to be ~ 2.9 K, whereas if we had neglected the ice-albedo feedback the temperature increase would have been only ~ 2.0 K.

Various estimates have been made of the feedback effects likely to be caused by changes in cloud amount and cloud type. These estimates range from λ_{cloud} is zero (i.e. the effects cancel) to recent GCM results which suggest that λ_{cloud} could be as large as -0.8 W m^{-2} K^{-1}. The addition of this feedback effect to those considered above would raise the surface temperature increase due to doubling CO_2 to ~ 6.5 K. This example demonstrates clearly how powerful a combination of positive feedback effects can be for the predicted surface temperature change.

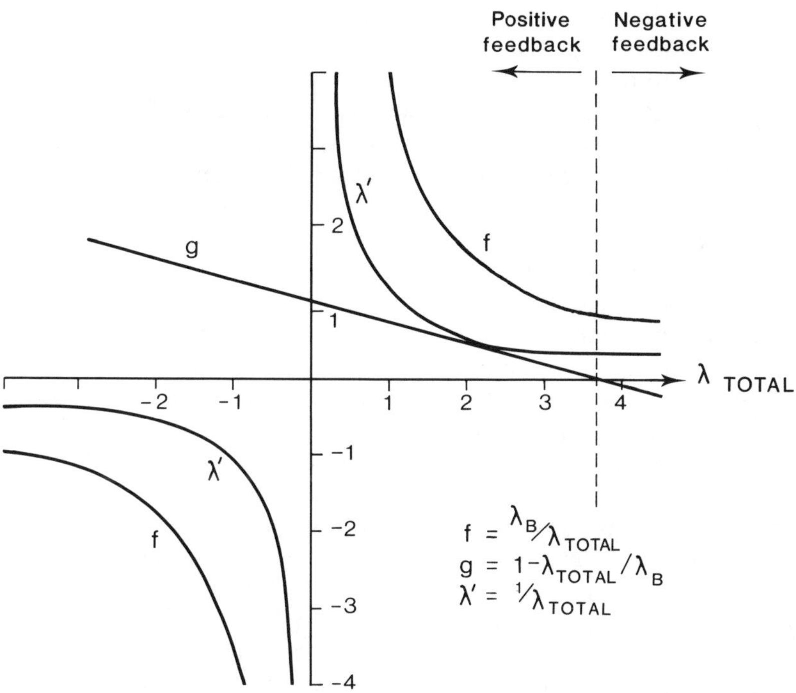

Figure 1.9 The feedback factor, f, the gain of the same system, g, and a third climate system feedback parameter λ', all plotted as a function of λ_{TOTAL} (probably the most useful measure of climate system sensitivity), the sum of all the contributing λ_is; λ_B is the value of λ_{TOTAL} for zero feedback. The areas of the diagram signifying overall positive and negative climatic feedback are shown. (The older sensitivity parameter to be found in the literature, the β parameter, not shown here, can be written as $\beta = S_0/\lambda_{TOTAL}$, where S_0 is the global average incoming solar radiation ($\simeq 340$ W m^{-2}: one fourth of the solar constant). Consequently for $\lambda_{TOTAL} = 3.75$ W m^{-2} K^{-1}, $\beta = 91.33$ K.)

Great care must be taken in interpretation of quoted values for feedback factors since several different definitions are in common use. Three of these are shown in Figure 1.9 plotted as functions of the feedback factor, λ, used here. The feedback factor used in the figure is λ_{TOTAL} as defined in equation (1.6). The value $\lambda_{TOTAL} = 3.75$ W m^{-2} K^{-1} represents that system with zero feedback. The values of the other feedback parameters at this point can be read off the graph. Note the areas of the graph (and the values of the feedback factors) which represent positive and negative feedback. As the term λ is often used synonymously with λ_{TOTAL} it is important to establish which is meant by careful reading of the context.

The climatic system is clearly extremely heterogeneous. There are many subsystems which interact with one another producing feedback effects. Climate dynamics is not unique in being controlled by changing feedback effects. If we reconsider the feedbacks affecting the population of a city, mentioned earlier, it is easy to imagine a range of other feedbacks operating in the same, and opposite, direction as the negative feedback on population induced by the perception of declining environmental character. As the city grows there are greater profits to be made in centrally located businesses (a positive feedback on immigration) while land prices, rents etc. increase (a negative feedback), muggings probably increase but long journey-to-work times are detrimental to family life. All these and many other feedback effects operate in a dynamically changing 'control' of the city size. Climatic feedbacks can be thought of as analogous to these geographical and economic controls.

Often the importance of feedback effects depends upon the timescale of behaviour of the subsystems they affect. This concept of timescale of response is crucially important in all branches of climate modelling. This timescale is variously referred to as the equilibration time, the response time, the relaxation time or the adjustment time. It is a measure of the time the subsystem takes to re-equilibrate following a small perturbation to it. A short equilibration timescale indicates that the subsystem responds very quickly to perturbations and can therefore be viewed as being quasi-instantaneously equilibrated with an adjacent subsystem which possesses a much longer equilibration time. It is common to express equilibration times in terms of the time, termed the *e*-folding time, it would take a system or subsystem to reduce an imposed displacement to $1/e$ th of the displaced value. For example, a pot of water removed from a stove will re-equilibrate with the room environment with an *e*-folding time depending upon the difference in temperature of the pot contents and the room, and the size and shape of the pot. A smaller temperature difference, a smaller pot or a larger surface-to-volume ratio of the container will result in relatively shorter *e*-folding times. Large *e*-folding times are possessed by subsystems which respond only very slowly. The response time is generally assessed in terms of the thermal response time. Table 1.1 lists equilibration times for a range of subsystems of the climate system. The longest equilibration times are those for the deep ocean, and the glaciers and ice sheets ($\sim 10^{10}$–10^{12} seconds); while the remaining elements of the climate system have equilibration times nearer 10^5–10^7 seconds.

Table 1.1 Equilibration times for several subsystems of the climate system

Climatic domain	Equilibration time	
	Seconds	Equivalent
Atmosphere		
Free	$\sim 10^6$	~11 days
Boundary layer	$\sim 10^5$	~24 hours
Ocean		
Mixed layer	$\sim 10^8 - 10^9$	~7–8 years
Deep	$\sim 10^{10}$	~300 years
Sea-ice	$\sim 10^6 - 10^{10}$	Days–100s of years
Continents		
Lakes and rivers	$\sim 10^6$	~11 days
Soil/vegetation	$\sim 10^6$	~11 days
Snow and surface ice layer	$\sim 10^5$	~24 hours
Mountain glaciers	$\sim 10^{10}$	~300 years
Ice sheets	$\sim 10^{12}$	~3 000 years
Earth's mantle	$\sim 10^{15}$	~30 million years

Thus the climatic system can be pictured as in Figure 1.10 in terms not only of subsystems and their directions and types of interactions, but also in terms of their approximate equilibration times. The very long equilibration time of the deep ocean poses a particularly difficult problem for climate modellers (cf. the fourth model described in Appendix B). The methods by which the short response time of the atmospheric features can be linked to the much slower

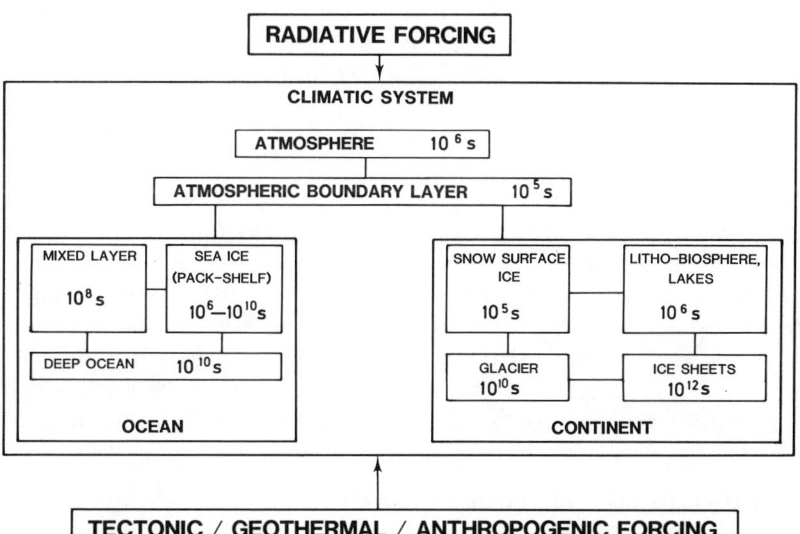

Figure 1.10 Schematic representation of the domains of the climatic system showing estimated equilibration times (reproduced by permission of Academic Press from Saltzman, 1983).

response time ocean system are discussed in Chapter 6. It must be noted, however, that some elements of the cryosphere have equally long response times and have, so far, not been included by parameterization into GCMs.

Clearly, modelling of climatic feedbacks will be crucial to the results of the modelling experiment. The exclusion of a feedback mechanism could radically alter results. Understanding feedbacks can only come through careful examination of the action of likely perturbations and the relative equilibration times of various parts of the climate system. Such a wide range of timescales is reflected in the wide range of model types currently in use.

1.5 PERTURBATIONS ON THE CLIMATE SYSTEM

The climate system is a dynamic system in transient balance. This concept, which is vitally important in climate modelling, is easy to visualize in terms of vehicle movement. The heart of New York City, Manhattan Island, suffers (or enjoys) a very large vehicular influx each morning and an equally large outflux in the evening. Over time periods greater than a few days Manhattan has an (approximate) vehicular balance, while over time periods of a few hours there are large negative and positive fluxes of vehicles. If the authorities were either (i) to close all bridges and tunnels on only the east side of the island, or (ii) to close all the car parks and refuse to allow street parking, the fluxes of vehicles would alter considerably and the net flux budget would change in this part of the New York subsystem.

Fluxes are thus seen to be vectors (they are the movement of some quantity, from somewhere to somewhere else, and the direction of flow is important) and net fluxes differ considerably as a function of the time period considered. Also different budgets, the result of the net fluxes, are established when the imposed disturbance changes. The most important fluxes in the climate system are fluxes of radiant (solar and heat) energy, but fluxes of water and mass will also be found to affect climate dynamics.

Perturbations to the climate system may be considered as falling into two separate categories. External perturbations are caused by variations in agents outside the climate system such as solar radiation fluctuations. On the other hand internal perturbations, such as volcanic eruptions, ice-sheet changes, CO_2 increases and deforestation, are variations in components of the climate system. Longer-term internal perturbations as a result of continental drift and mountain building have an effect, and changes in the polarity of the Earth's magnetic field may influence the upper atmosphere and thus, perhaps, the whole climate.

1.5.1 External causes of climatic change

Milankovitch variations

The astronomical theory of climate variations, also called the *Milankovitch theory*, is an attempt to relate climatic variations to the changing parameters

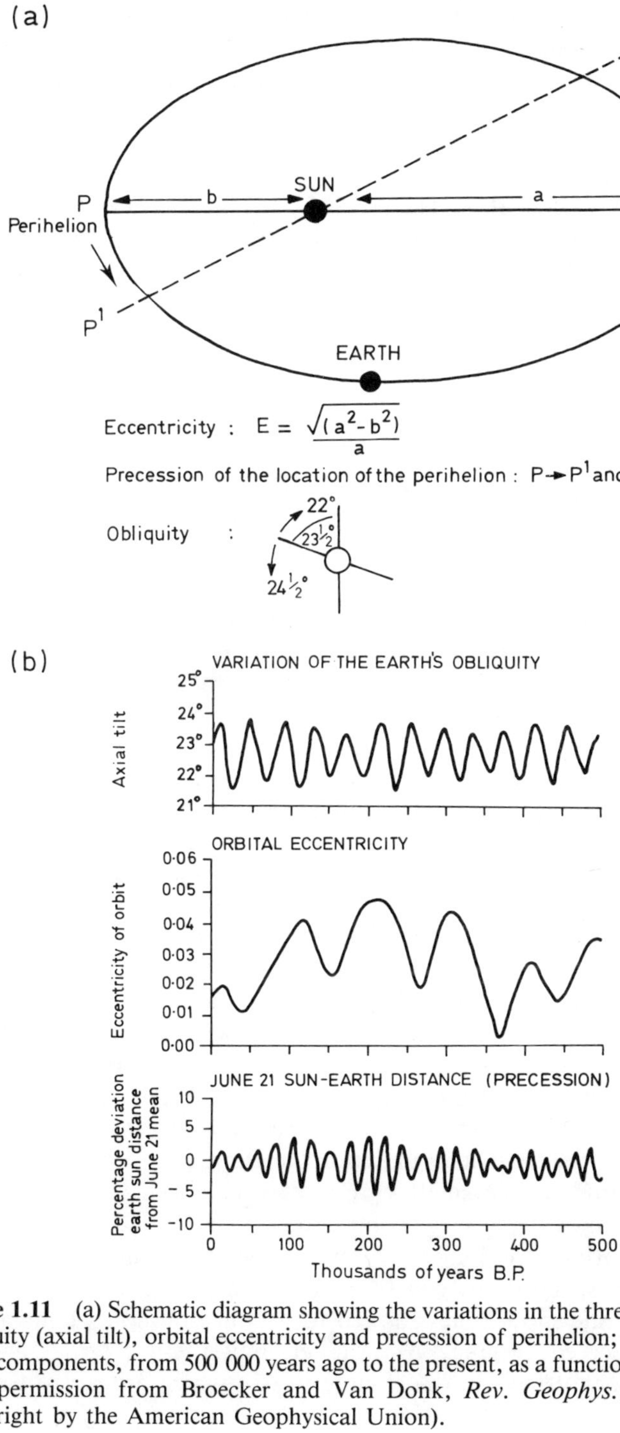

Figure 1.11 (a) Schematic diagram showing the variations in the three orbital components: obliquity (axial tilt), orbital eccentricity and precession of perihelion; (b) variations in these three components, from 500 000 years ago to the present, as a function of time (reproduced with permission from Broecker and Van Donk, *Rev. Geophys.*, **8**, 169–196 (1970). Copyright by the American Geophysical Union).

of the Earth's orbit around the Sun. There are several different ways in which the orbital configuration can affect the received radiation and thus possibly the climate. They are (Figure 1.11): (i) *Changes in eccentricity.* The Earth's orbit becomes more circular and then more elliptical in a pseudo-cyclic way, completing the cycle in about 110 000 years. The mean annual incident flux varies as a function of the eccentricity of the orbit, E. For a larger value of E there is a smaller incident annual flux. The current value of E is 0.017. In the last 5 million years it has varied from 0.000 483 to 0.060 791. These variations would result in changes in the incident flux of +0.014 to −0.17 per cent from the current value. (ii) *Changes in obliquity.* The obliquity, the tilt of the Earth's axis, is the angle between the Earth's axis and the plane of the ecliptic (the plane in which the bodies of the solar system lie). This tilt varies from about 22° to 24.5°, with a period of about 40 000 years. The current value is 23.5°. Seasonal variations depend upon the obliquity: if the obliquity is large so is the range of seasonality. However, the total received radiation is not altered, but a greater seasonal variation in received flux is accompanied by a smaller meridional gradient in the annual radiation. (iii) *Orbital precession.* The orbit of the Earth is an ellipse around the Sun, which lies at one of the foci. Owing to gravitational interaction with the other planets, primarily Jupiter, the perihelion (the point of the Earth's orbit closest to the Sun) moves in space so that the ellipse is moved around in space. This precession will cause a progressive change in the time of the equinoxes. These changes occur in such a way that two periodicities are apparent: 23 000 years and 18 800 years. This change, like that of obliquity, does not alter the total radiation received but does affect its temporal and spatial distribution. For example, perihelion is currently on 5 January, in the middle of northern hemisphere winter, but 11 000–15 000 years from now it will occur in July. At the present-day value of eccentricity there is a range of ~6 per cent in the solar constant between perihelion and aphelion (i.e. 1411 to 1329 W m^{-2}).

Modelling results have suggested that the present configuration of the land masses in the northern hemisphere may favour rapid development of ice caps when conditions favour cool northern hemisphere summers.

Spectral analysis of long-term temperature data has shown the existence of cycles with periods of ~20 000, ~40 000 and ~100 000 years (Figure 1.12). These correspond closely with the Milankovitch cycles. The strongest signal in the observational data, however, is the 100 000-year cycle. This cycle corresponds to that of eccentricity variations in the Earth's orbit but eccentricity variations produce the smallest insolation changes. Hence the exact effect of the Milankovitch cycles is far from clear. While they offer an interesting 'explanation' for long-term, cyclic climatic changes, the energy distributions within spectral analyses of climate and of orbital variations are radically different and models have failed to produce observed temperature changes from observed forcing. Almost certainly these external changes trigger large feedback effects in the climate system which are yet to be fully understood.

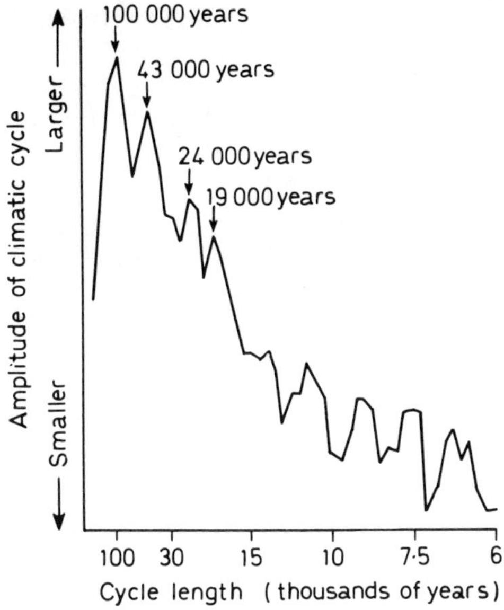

Figure 1.12 Spectrum of climatic variations over the past 500 000 years. The graph shows the importance of the climatic cycles of 100 000 years (eccentricity); 43 000 years (obliquity) and 24 000 and 19 000 years (precession of the location of the perihelion). The curve is constructed from an isotopic record of two Indian Ocean cores (from Imbrie and Imbrie (1979) by permission of Macmillan, London and Basingstoke).

Solar activity

Variations in the climate during historical times have been linked with the sunspot cycle, which is a second possible cause of solar-produced climatic change. This cycle occurs with a 22-year periodicity: the 'Hale' double sunspot cycle. The overall amplitude of the cycles seems to increase slowly and then fall rapidly with a period of 80–100 years. There also appears to be a quasi-cyclic fluctuation of the order of 180 years. No mechanistic link between sunspot activity and surface conditions on the Earth has been demonstrated. Correlations between climate and sunspots usually fail when global conditions are considered. In particular, the Little Ice Age has been linked to the 'Maunder minimum' in sunspots; although it should be noted that the actual period of the Little Ice Age itself seems to vary according to the geographical area from which data are taken.

Other external factors

Collisions of comets with the Earth and very large meteoritic impacts have also been proposed as causes of climatic fluctuations. Many of the disturbances that such impacts would cause, such as an increase in stratospheric and tropospheric aerosols, are similar to disturbances internal to the system, described in the next subsection.

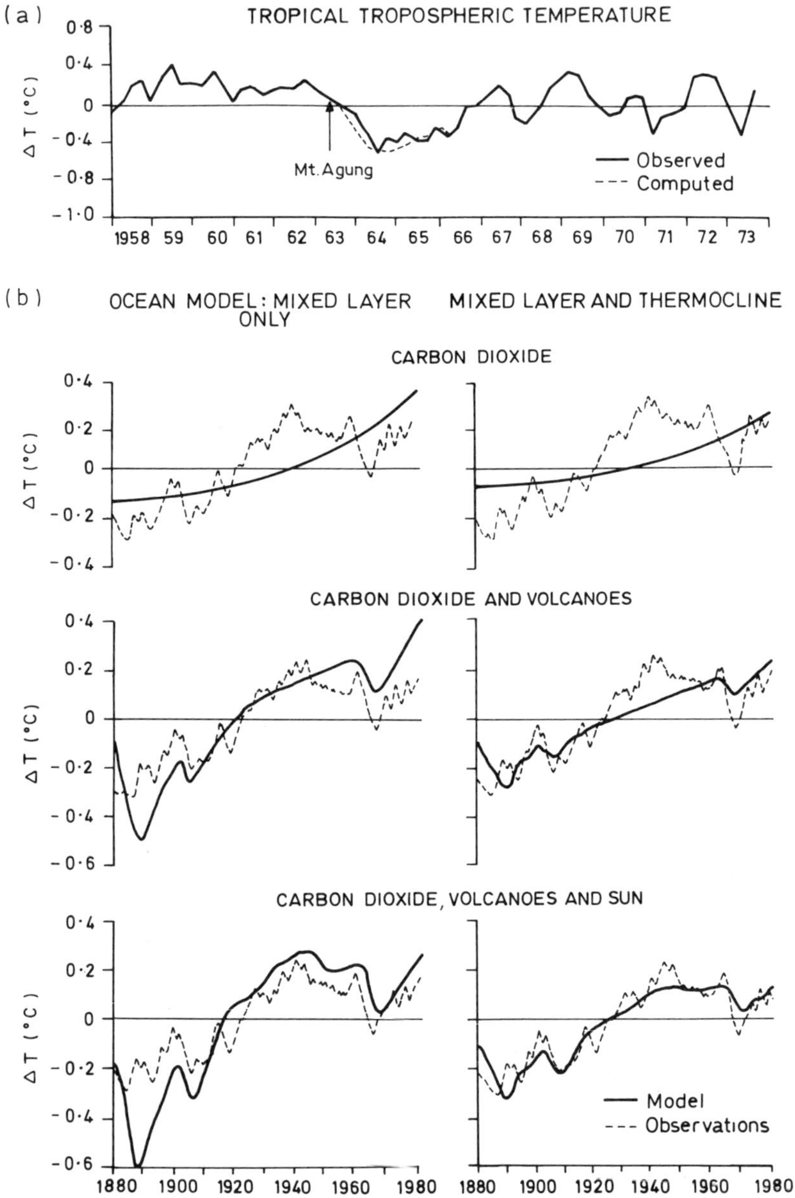

Figure 1.13 (a) Observed tropospheric temperatures for the region between 30° N and 30° S and model-computed temperatures for the period after the eruption of Mount Agung in 1963. Both computed and observed temperatures exhibit a significant decrease as a result of the volcanic eruption (after Hansen *et al.*, 1980); (b) global temperature trends obtained from a one-dimensional RC climate model, taking changes in atmospheric CO_2, volcanic activity and solar luminosity into account. The results are in good agreement with the observed surface temperatures (reproduced with permission from Hansen *et al.*, *Science*, **213**, 957–966. Copyright 1981 by the AAAS).

1.5.2 Internal causes of climatic change

A change in an internal component of the climate system will impose a climatic perturbation. Most current concern is about the possible effects of man, which could operate on the relatively short timescales necessary to create noticeable changes within the next century. The only natural effect which is thought likely to be important on this timescale is volcanic activity.

Volcanic eruptions

Volcanoes influence climate by projecting large quantities of particulates and gases into the atmosphere. Volcanic eruptions certainly produce measurable temperature anomalies of at least a few tenths of a degree (Figure 1.13(a)). The effect of the injected aerosol upon the radiation balance and whether heating or cooling ensues will depend largely on the height of injection into the atmosphere.

Most eruptions inject particulates into the troposphere at heights between 5 and 8 km. These are rapidly removed either by gravitational fall-out or rain-out and the resultant climatic effect is minimal. More violent eruptions hurl debris into the upper troposphere or even into the lower stratosphere (15–25 km) (e.g. Mount Agung in 1963 and El Chichón in 1982). Eruptions such as these are much less frequent but are likely to have more extensive climatic effects. The particulates have a long residence time in the stratosphere: of the order of a year for aerosols of radii 2–5 μm but as long as 12 years for smaller aeorosols of radii 0.5–1.0 μm. Thus they can become widely distributed by the stratospheric circulation. Non-absorbing aerosols increase the albedo of the atmosphere and reduce the amount of solar radiation that reaches the surface. If the aerosol absorbs in the visible part of the spectrum, energy is transferred directly to the atmosphere. If the aerosol absorbs and emits in the infrared, the greenhouse effect is increased.

Immediately following an eruption the stratosphere is dominated by particles which scatter radiation of wavelengths up to 10 μm roughly ten times as efficiently as normal stratospheric particles. The 'clear sky' optical thickness can rise to 0.1 (20 times the normal value) after large eruptions since the particles also absorb visible radiation. Sulphate production is increased a few months later and a further increase in the visible scattering occurs along with a slight increase in the infrared absorption. These changes will affect the atmospheric heating rates. The enhanced absorption of visible radiation is not sufficient to compensate for enhanced cooling by emission of infrared radiation.

Model simulations suggest that radiative effects will, overall, produce a global cooling when large scale volcanic eruptions occur. Figure 1.13(b) shows results of a simulation in which an attempt was made to reproduce recent temperature changes considering changes in solar luminosity and atmospheric CO_2 in addition to volcanoes. Volcanoes were included via a 'dust veil index' developed from a historical analysis catalogue of volcanic eruptions. The results support the idea that volcanoes have a cooling effect, but complications arise because these volcanic particles can serve as cloud condensation nuclei. Consequently feedbacks which involve cloudiness changes may be invoked. However, eruptions and their effects on the atmosphere are very short-lived compared to the time needed to influence the heat storage of the oceans. Hence temperature anomalies do not persist, nor are they likely to initiate significant long-term climatic changes.

Tropospheric particulates and clouds

There is no direct evidence of any increase in particulate (aerosol) loading of the atmosphere and many measurements suggest that only following large volcanic eruptions does the atmospheric loading change. Apparently such particles can be neglected as a possible climatic change mechanism. An increase in the electrical conductivity of the atmosphere observed over the past 70 years may be caused by more aerosols in the size range 0.02–0.2 μm which are likely to be attributable to the activities of man rather than any natural source. Again, this change may lead to variations in both radiative exchanges and cloud amount with associated climatic implications.

Cloud variations, both of type and amount, may result from many different processes. One possible influence is the abundance of cloud condensation nuclei.

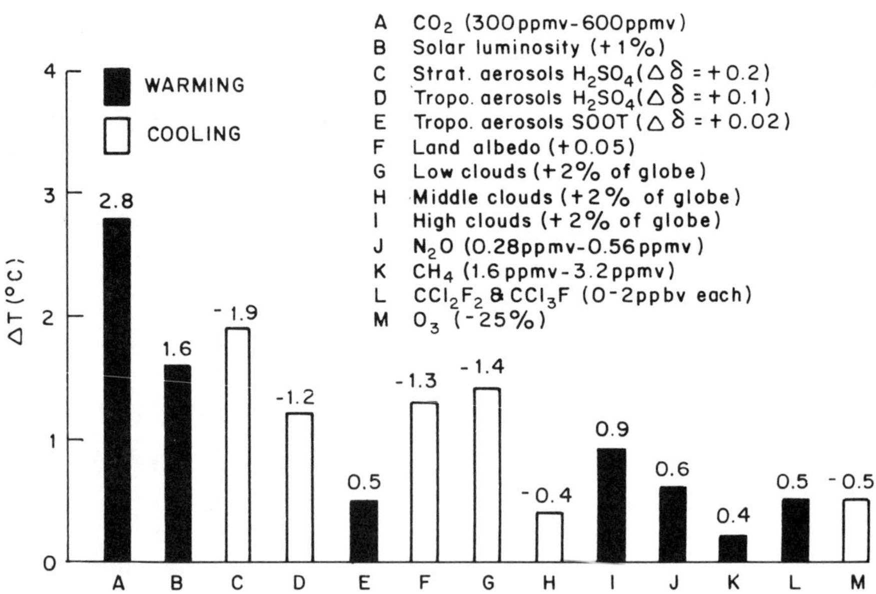

Figure 1.14 The effect of various climatological perturbations upon surface temperatures as predicted by a one-dimensional RC climate model. One of the major perturbations causing cooling is seen to be increased loading of stratospheric aerosols. Column I shows that increasing cirrus cloud operates to warm, rather than to cool as do increases in middle- and low-level clouds (columns G and H). Warming by a 2 per cent increase in high clouds is greater than the cooling caused by a 2 per cent increase in middle-level clouds (reproduced with permission from Hansen *et al.*, *Science*, **218**, 957–966. Copyright 1981 by the AAAS).

Aerosols injected into the lower troposphere tend to be fairly rapidly removed. Although this probably means an increase in total condensation and rainfall, the effects are localized and of short duration. Input into the stratosphere is unlikely to affect cloudiness because water vapour is unable to penetrate into the stratosphere to any great extent. Injections into the upper troposphere may have far-reaching effects since at heights of 10–18 km these particles, whether volcanic debris or of human origin, could increase the amount of ice clouds. A change in cirrus cloud cover has recently been described as one of the possibly critical climatic perturbing agents (Figure 1.14), since an increase in cirrus cloud amount will increase absorption of the infrared radiation without a balancing increase in reflection of shortwave radiation leading to a general climatic warming. Figure 1.14 illustrates the relative magnitude of several different climate perturbations. The results were derived from a one-dimensional RC model.

Stratospheric ozone and tropospheric carbon dioxide

Gaseous pollution emissions are believed to perturb the ozone layer of the lower stratosphere. The prime concern is with chlorofluorocarbons commonly called 'freons', which are used as refrigerants and spray-can propellants. It is by no means clear whether the net result of the interaction between ozone and chlorofluorocarbons will be an increase or a decrease in stratospheric ozone or whether the change will be climatically significant. Chlorofluorocarbons which remain in the troposphere have recently been shown to possess infrared absorption features which means that they will contribute increasingly to the greenhouse effect as their concentration increases.

The inclusion of chemistry within most climate models has not yet been attempted, but the effect of chemical perturbations can be investigated by imposing on a model the supposed gaseous concentrations after a chemical change.

The discovery, in 1986, of 'holes' in the ozone layer in the high latitudes of both hemispheres focussed attention upon the urgent need for interactive chemical submodelling in climate models. As yet it is unclear whether these regions of decreased stratospheric ozone levels are attributable to anthropogenerated disturbance or unusual solar activity.

The increasing quantity of carbon dioxide in the atmosphere has been well documented and modelling of its effect is widespread in the climate-modelling literature. The immediate effect of this global increase in atmospheric CO_2 will be to raise the surface temperature by the 'greenhouse' effect (Figure 1.14). Although CO_2 is itself a greenhouse gas the majority of the increased greenhouse effect induced by CO_2 is due to increased temperatures permitting a larger concentration of atmospheric water vapour which is a still more efficient greenhouse absorber. The magnitude of the temperature increase will depend on the nature of the interaction between all involved feedback mechanisms.

Surface changes

Mankind is beginning to make what can be regarded as regional-scale changes to the character of the Earth's surface. Climate modellers have also investigated the climatic effect of changes in the nature of the Earth's surface. (A simple model of the possible impact on the climate of anthropogenerated surface albedo change is described in Appendix B.)

Desertification is a problem which affects millions of people and the environmental threat of deforestation continues. Adapting global-scale models to regional conditions by allowing them to emphasize those processes, such as cloud amount or surface conditions, deemed important for a particular region, permits regional 'predictions'.

(i) *Desertification.* The sparse vegetation natural to arid and semi-arid areas can be easily removed as a result of relatively minor changes in the climate or by overgrazing. Removal of vegetation and exposure of bare soil decrease soil water storage, because of increased runoff, and increase albedo. Less moisture available at the surface means decreased latent heat flux leading to an increase in surface temperature. On the other hand the increased albedo produces a net radiative loss. In climate model calculations the latter effect appears to dominate and the radiation deficit causes large scale subsidence. In this descending air, cloud and precipitation formation would be impossible and aridity would increase. The result of a relevant model simulation is shown in Figure 1.15(a). This global simulation involves a surface albedo change for a group of semi-arid areas. It can be seen that an increase in surface albedo does seem to decrease rainfall. Use of a global model emphasizes that all parts of the climate system are interlinked. Although this particular model includes many simplifications, the results are illustrative of the types of surface-induced climatic effects which are possible.

(ii) *Deforestation.* Modellers have also attempted to examine the climatic effects of forest clearance, since whenever land is cleared for agricultural purposes there is a marked change in surface character. This change can be especially noticeable when forests are replaced by cropland. At present around 30 per cent of the land surface of the Earth is forested and about a third as much is cultivated. However, the amount of forest land, particularly in the tropics, is rapidly being reduced and the surface characteristics of large areas are being irreversibly changed. One area which is undergoing deforestation is the Amazon Basin in Brazil. The important change in deforestation is in the surface hydrological characteristics since the evapotranspiration from a forested area can be many times greater than from adjacent open ground. In one climate model simulation the tropical forest vegetation over an area of 5 million km^2 of the Amazon Basin was converted to impoverished grassland: a massive alteration which could occur within 30–60 years at the present rate of deforestation. At the end of the experiment it was found that soil moisture and evaporation had both decreased (Figure 1.15(b)). The surface temperature increased by around 3 K, despite the compensation for the decreased latent heat flux away from the surface by the increase in albedo. The changes were a regional-scale phenomenon, but no global-scale changes were detected. However, the model

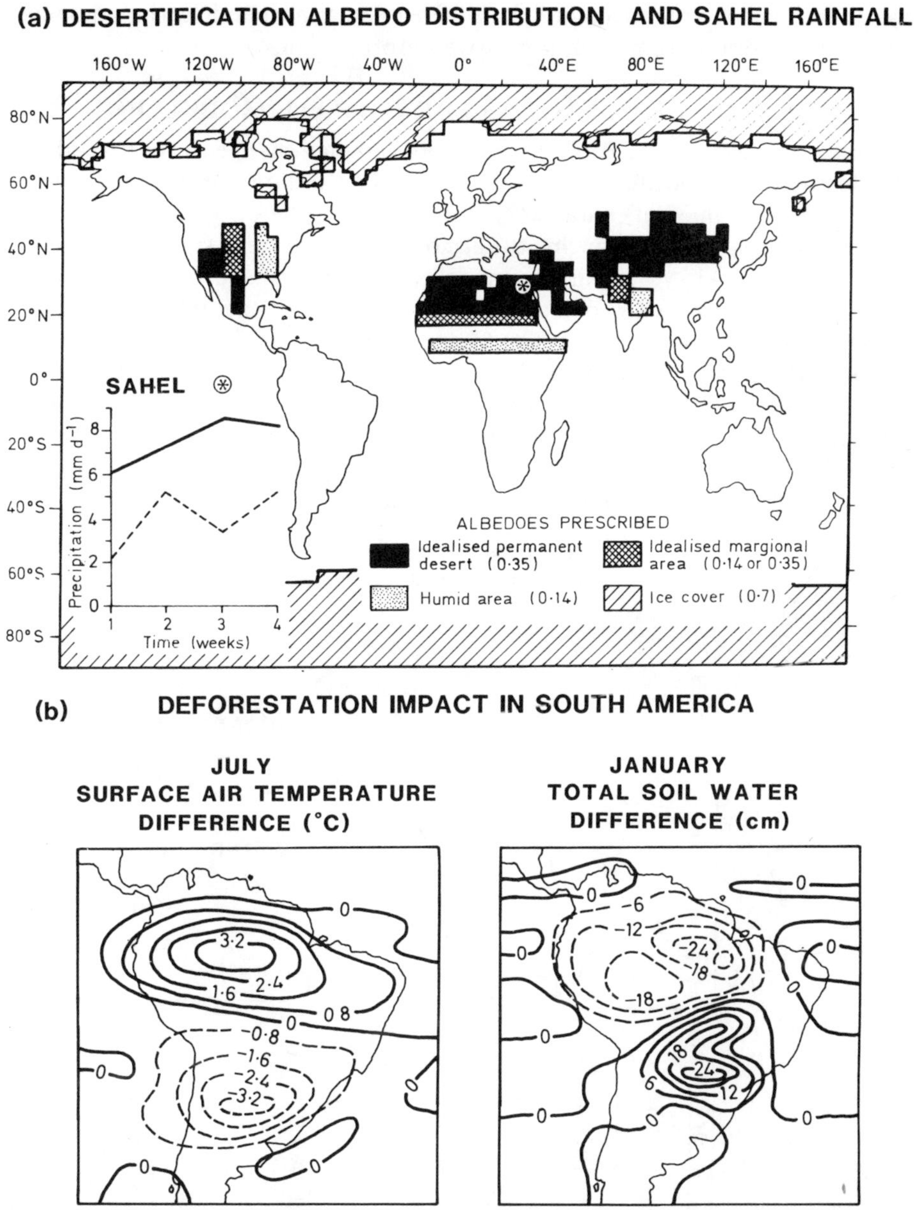

Figure 1.15 (a) The distribution of areas for which albedo changes were made in a set of climate model experiments, originally conducted by Charney (1975), designed to examine desertification. The inset graph shows the rainfall resulting from increasing the surface albedo from 0.14 to 0.35 in the Sahel region when free evaporation was permitted (redrawn with permission from Henderson-Sellers and Wilson, *Rev. Geophys. Space Phys.*, **21**, 1743–1778 (1983). Copyright by the American Geophysical Union). (b) Simulated surface air temperature and soil moisture differences in July and January respectively following the replacement of the Amazon forest by impoverished grassland.

was not designed to take account of the increase in atmospheric carbon dioxide which would result from such a massive deforestation.

1.6 RANGE OF QUESTIONS FOR CLIMATE MODELLING

The type of question asked of climate modellers has changed over the three-decade history of modelling. In the early years the question was 'Can the model capture the fundamental characteristics of the atmosphere?' so that particular attention was paid to reproducing the atmospheric time mean state adequately. Over the last 10–15 years climate models of all types have been applied to questions which are strictly demands for predictions: for example 'What is the impact of doubling atmospheric CO_2?' and 'Will removing the tropical forests of the world affect the climate at locations distant from the deforested area?' As will become clear in the rest of this book, many of these 'predictions' have been made with models that have not yet been fully tested. Sensitivity testing and 'validation' of climate models has only become widespread since the models have been shown to be doing a good basic job.

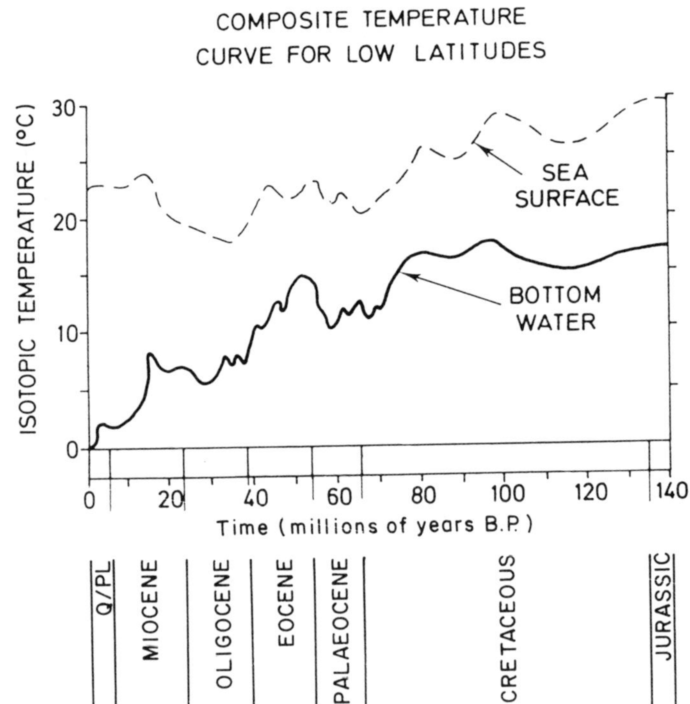

Figure 1.16 Compilation of oxygen isotope palaeotemperature data obtained by analysis of benthic and planktonic *foraminifera* from Deep Sea Drilling Project cores. [Q/PL denotes the Pliocene epoch and the Quaternary period, the latter beginning about 2 million years before present (B.P.).] The upper curve is drawn through tropical sea-surface temperatures and the lower curve through bottom water temperatures (reproduced with permission from Douglas and Woodruff, 1981).

The next stage in the evolution of climate modelling seems likely to be an attempt to answer still more difficult questions about the climate system. One such question which could not even be considered by modellers until they had coupled an ocean sub-model with an atmospheric model is what was the cause of the very high ocean bottom-water temperatures in the Cretaceous period (60–130 million years ago). Figure 1.16 shows a recent compilation of tropical sea-surface and bottom-water temperatures. The tropical sea-surface temperatures vary very little over the 140 million years shown while the bottom-water temperatures have decreased by more than 10 K since the Cretaceous period. How these high, deep ocean temperatures could have occurred and what effects the changes which caused them had on other components of the climate are questions which models can, perhaps, be used to answer. The other question which is of current importance for climatologists is 'What is the direction and magnitude of cloud feedback?' Modellers cannot yet hope to offer a complete answer to this searching question, but model simulations and the continuing search for better parameterization schemes prompt and focus observational efforts.

The rest of this book is intended to give the reader a basic understanding of the types and complexity of climate models. We have not tried to answer specific questions such as those outlined above or describe, in detail, particular models or experiments. In writing this primer our aim has been to help those new to climate modelling to a quicker and fuller understanding of the available literature.

RECOMMENDED READING

Dickinson, R. E., (1985) Climate sensitivity. In S. Manabe (ed.) *Issues in Atmospheric and Oceanic Modelling, Part A, Climate Dynamics*, Advances in Geophysics, Vol. 28, Academic Press, New York, pp. 99–129.

GARP (1975) The physical basis of climate and climate modelling, *GARP Publication Series No. 16*, WMO/ICSU, Geneva.

Hansen, J. E., and Takahashi, T. (eds) (1984) *Climate Processes and Climate Sensitivity*, Geophysical Monograph 29, Maurice Ewing Vol. 5, American Geophysical Union, Washington DC, 368pp.

Houghton, J. T. (ed.) (1984) *The Global Climate*, Cambridge University Press, Cambridge, 233pp.

Manabe, S. (ed.) (1985) *Issues in Atmospheric and Oceanic Modelling, Part A, Climate Dynamics*, Advances in Geophysics, Vol. 28, Academic Press, New York, 591pp.

National Research Council (1982) *Carbon Dioxide and Climate: A Second Assessment: Report of the CO_2/Climate Review Panel*, National Academy Press, Washington DC.

Oort, A. H., and Peixóto, J. P. (1983) Global angular momentum and energy balance requirements from observations, *Adv. Geophys.*, **25**, 355–490.

Peixóto, J. P., and Oort, A. H. (1984) Physics of climate, *Rev. Mod. Phys.*, **56**, 365–430.

Saltzman, B. (ed.) (1983) *Theory of Climate*, Advances in Geophysics, Vol. 25, Academic Press, New York, 505pp.

Schlesinger, M. E. (ed.) (1987) *Physically Based Climate Models and Climate Modelling*, Proceedings of a NATO ASI, Reidel, Dordrecht.

Chapter 2

A History of and Introduction to Climate Models

'There has been a renaissance in climate studies over the last decade. Scientists in different disciplines concerned with the climate system have grown increasingly appreciative of the connections between the various components of the climate system and the hazards of overly narrow viewpoints.'
R. E. Dickinson (1983)

2.1 INTRODUCING CLIMATE MODELLING

Any climate model is an attempt to simulate the many processes that produce climate. The objective is to understand these processes and to predict the effects of changes and interactions. The simulation is accomplished by describing the climate system in terms of basic physical laws. Hence a model can be considered as being comprised of a series of equations expressing these laws. Climate models can be slow and costly to use, even on the fastest computer, and the results can only be approximations.

The need for simplification

For several reasons a model must be a simplification of the real world. The processes of the climate system are not fully understood, but they are known to be complex. Furthermore, they interact with each other, producing *feedbacks* (Section 1.4), so that any solution of the governing equations must involve a great deal of computation. The solutions that are produced start from some initialized state and investigate the effects of changes in a particular component of the climate system. The boundary conditions, for example the solar radiation or sea-surface temperatures, are set from observational data but these data are rarely complete or of adequate accuracy to specify completely the environmental conditions, so that there is inherent uncertainty in the results.

Large scale global climate models, designed to simulate the climate of the planet, must take into account the whole climate system (see Figure 1.2). All of the interactions between components must be integrated in order to develop a climate model. This presents great problems, because the various interactions operate on different timescales. For example, the effects of deep ocean overturning may be very important when considering climate averaged over decades to centuries, while local changes in wind direction may be unimportant on this timescale. If, on the other hand, monthly timescales are of concern, the relative importance would be reversed.

Global-scale models are usually concerned with long-term average conditions. It is common, for example, to develop models for average conditions in January and July. This is not to imply that a particular January in the period for which a climate model prediction is made will have these conditions, only that the conditions apply to an average January. Indeed, it is always implied that any 'new' climate predicted will have variation about the mean, just as with the present climate. The amount of this new variation is often of concern when the results of global-scale models are used to estimate the possible impact of climatic change in a local or regional area.

The simplifications that must be made to the laws governing climatic processes can be approached in several ways. Consequently there are numerous different global-scale climate models available. In general, two sets of simplifications need to be made. The first involves the processes themselves. It is usually possible to treat in detail some of the processes, specifying their governing equations fully. However, other processes must be treated in an approximate way, either because of our lack of understanding or because there are inadequate computer resources to deal with them. For example, it might be decided to treat the radiation processes in great detail, but only approximate the horizontal energy flows associated with regional-scale winds. The approximation may be approached either by using available observational data, the *empirical* approach, or through specification of the physical laws involved, the *theoretical* or *conceptual* approach.

Resolution in time and space

The second set of simplifications involves the resolution of the model in both time and space. While generally the finer the spatial resolution, the more reliable the results, constraints of both data availability and computational time may dictate that a model may have to have, for example, latitudinally averaged values as the basic input. In addition, too fine a resolution may be inappropriate because processes acting on a smaller scale than the model is designed to resolve may be inadvertently incorporated. Similar considerations are involved in the choice of temporal resolution. Most computational procedures require a 'timestep' approach to calculations. The processes are allowed to act for a certain length of time and the new conditions are calculated. The process is then repeated using these new values. This continues until the conditions at the required time have been established. Again, although accuracy potentially increases as the timestep decreases, there are constraints imposed by data, computational ability and the design of the model.

Although models are designed to aid in predicting future climates, their performance can only be tested against the past or present climate. Usually when a model is developed an initial objective is to ascertain how well it compares with the present climate. Thereafter it may be used to simulate past climates, not only to see how well it performs but also to gain insight into the causes of these climates. Although such past climates are by no means well known, this comparison provides a very useful step in establishing the validity of the modelling approach. After such tests, the model may be used to gain insight into possible future climates.

2.2 TYPES OF CLIMATE MODELS

The important components to be considered in constructing or understanding a model of the climate system are:

1. *Radiation*—the way in which the input and absorption of solar radiation and the emission of infrared radiation are handled.
2. *Dynamics*—the movement of energy around the globe (specifically from low to high latitudes) and vertical movements (e.g. convection).
3. *Surface processes*—inclusion of land/ocean/ice and the resultant change in albedo, emissivity and surface–atmosphere energy interchanges.
4. *Resolution in both time and space*—the timestep of the model and the horizontal and vertical scales resolved.

The relative importance of these processes and the physical (as opposed to empirical) basis for parameterizations employed in their incorporation can be discussed using the 'climate-modelling pyramid' (Figure 2.1). The edges represent three of the basic elements of models, and complexity is shown increasing upwards. Around the base of the pyramid are the simpler climate models which incorporate only one primary process.

As indicated in Figure 2.1, there are four basic types of model:

1. Energy balance models (EBMs) are one-dimensional models predicting the variation of the surface (strictly the sea-level) temperature with latitude. Simplified relationships are used to calculate the terms contributing to the energy balance in each latitude zone.
2. One-dimensional radiative–convective (RC) models compute the vertical (usually globally averaged) temperature profile by explicit modelling of the radiative processes and a 'convective adjustment' which re-establishes a predetermined lapse rate.
3. Two-dimensional statistical dynamical (SD) models deal explicitly with surface processes and dynamics in a zonally averaged framework and have a vertically resolved atmosphere.
4. General circulation models (GCMs). The three dimensional nature of the atmosphere or ocean is incorporated. In these models an attempt is made to represent most physical processes believed to be important. The spatial resolution is, at present, too coarse for synoptic scale processes to be modelled correctly in the ocean case and this discrepancy between length and time scales

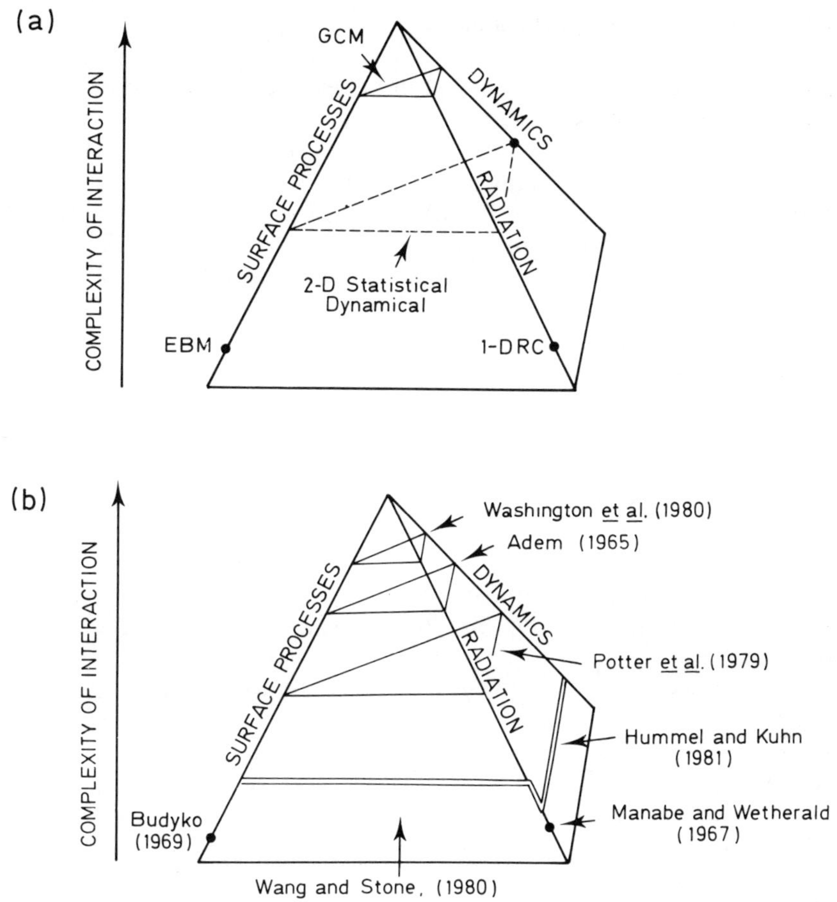

Figure 2.1 The climate-modelling pyramid. The position of a model on the pyramid indicates the complexity with which the three primary processes interact. The base of the pyramid can be considered hollow since there is essentially no interaction between the primary processes. Progression up the pyramid leads to greater interaction between each primary process. The vertical axis is not intended to be quantitative. (a) The positions of the four basic model types. (b) Particular climate models are seen to be based upon different methods of incorporating the primary processes and the level of complexity of the interactions (after Shine and Henderson-Sellers, 1983).

in the atmosphere and ocean is the fundamental problem obstructing the development of coupled models.

There is some ambiguity concerning the expansion of "GCM". Two possible terms are "global climate model" and "general circulation model". The latter could refer to a weather forecast model so that in climate studies "GCM" is understood to mean "*general circulation climate model*". A further distinction is often drawn between oceanic general circulation models and atmospheric general circulation models by terming them OGCMs and AGCMs.

2.2.1 Energy balance climate models

These models have been instrumental in increasing our understanding of the climate system and in the development of new parameterizations and methods of evaluating sensitivity for more complex and realistic models. This type of model can be readily programmed and implemented on most small computers and the inherent simplicity of EBMs combined with the ease of interpreting results make them ideal instructional tools.

Energy balance models are generally 'one-dimensional', the dimension in which they vary being latitude. Vertical variations are ignored and the models are used with surface temperature as the predicted variable. Since the energy balance is allowed to vary from latitude to latitude, a horizontal energy transfer term must be introduced, so that the basic equation for the energy balance at each latitude, ϕ, is

$$C_m \frac{\Delta T(\phi)}{\Delta t} = R{\downarrow}(\phi) - R{\uparrow}(\phi) + \text{transport into zone } \phi \tag{2.1}$$

where C_m is the heat capacity of the system and can be thought of as the system's 'thermal inertia' and $R{\downarrow}$ and $R{\uparrow}$ are the incoming and outgoing radiation fluxes respectively.

The radiation fluxes at the Earth's surface must be parameterized with care since conditions in the vertical are not considered in this type of model. To a large extent the effects of vertical temperature changes are treated implicitly. In a clear atmosphere, convective effects tend to ensure that the lapse rate remains fairly constant. However, cloud amount depends only weakly on surface temperature, so that cloud albedo is only partially incorporated in the model. In particular, clouds in regions of high temperatures, such as the intertropical convergence zone, are ignored in the parameterization of albedo.

When using the model for annual average calculations the surface albedo can be regarded as constant for a given latitude. This type of model, however, can also be used for seasonal calculations. In this case it is usual to allow the albedo to vary with temperature to simulate the effects of changes in sea-ice extent.

The dynamical events of the real world are not modelled in an EBM and it is assumed that a 'diffusion' approximation is adequate for including heat transport. This approximation relates energy flow directly to the latitudinal temperature gradient. This flow is usually expressed as being proportional to the deviation of the zonal temperature, T, from the global mean, $\bar{T}$.

Early models were originally found to be stable only for small perturbations away from present-day conditions. For instance, they predicted the existence of an ice-covered Earth for only slight reductions in the present solar constant. However, this result did give rise to studies of the sensitivity of various climate model types to perturbations (see Section 2.4).

2.2.2 One-dimensional radiative–convective climate models

One-dimensional RC models represent an alternative approach to relatively simple modelling of the climate and they also occur at the bottom of the modelling pyramid (Figure 2.1). In this case the 'one dimension' in the name refers to altitude. One-dimensional RC models are designed with an emphasis on the global average surface temperature, although temperatures at various levels in the atmosphere can be obtained. The models can also be applied to zonally averaged conditions, by including a description of the horizontal energy transport.

The main emphasis in these models is on the explicit calculation of the fluxes of solar and terrestrial radiation (the radiation streams). Given an initially isothermal atmosphere the heating rates for a number of layers in the atmosphere are calculated, although the cloud amount, optical properties and the albedo of the surface generally need to be specified. The temperature change in each layer which results from an imbalance between the net radiation at the top and bottom of the layer is calculated. At the end of each timestep a revised radiative temperature profile is produced. If the calculated lapse rate exceeds some predetermined 'critical' lapse rate the atmosphere is presumed to be convectively unstable. An amount of vertical mixing, sufficient to re-establish the prescribed lapse rate, is carried out and the model proceeds to calculate the next radiative timestep. This procedure continues until convective readjustment is no longer required and the net fluxes for each layer approach zero.

One-dimensional RC models operate under the constraints that at the top of the atmosphere there must be a balance of shortwave and longwave fluxes and that surface energy gained by radiation equals that lost by convection. However, they vary in the way they incorporate the critical lapse rate. Some use the dry adiabatic lapse rate, some the saturated one, while many use a value of $6.5\,\mathrm{K\,km^{-1}}$, which is the value in the US standard atmosphere (sometimes called the McClatchey atmosphere) profile. Similarly different humidity and cloud formulations are possible.

2.2.3 Two-dimensional climate models

Two-dimensional climate models can represent either the two horizontal dimensions or the vertical and one horizontal dimension. It is the latter which are more common, combining the latitudinal dimension of the energy balance models with the vertical one of the radiative–convective models. These models also include a more realistic parameterization of the latitudinal energy transports. In such models the general circulation is assumed to be composed mainly of a cellular flow between latitudes which is characterized using a combination of empirical and theoretical formulations. A set of statistics summarizes the wind speeds and directions while an eddy diffusion coefficient of the type used in the EBMs governs energy transport. As a consequence of this approach, these models are called 'statistical dynamical' (SD) models.

These models are about half-way up the modelling pyramid (Figure 2.1), being much more complicated than the vertically or latitudinally resolved one-dimensional models. Their use has provided insight into the operation of the present climate system, in particular showing that the relatively simple diffusion coefficient approach for poleward energy transports is appropriate, provided that the coefficient, as well as the transport, is allowed to vary with the latitudinal temperature gradient. Similarly, advances in the understanding of baroclinic waves have been achieved from studies of the results of these SD models. Despite this, these two-dimensional models are still rather limited in their applicability to climate prediction studies. They are insensitive to changes within a latitude band and changes such as new land–sea temperature contrasts and cloud regimes which might result from alterations of land-surface albedo are impossible to investigate although some compromise may be obtained by considering each zone as being divided into a land and ocean part. This type of 'two-channel' approach is discussed with reference to a more complex EBM in Section 5.6. As a result of this lack of zonal resolution, these two-dimensional models have largely been superseded by GCMs for consideration of the effect of perturbations on the present climate.

2.2.4 General circulation climate models

The aim of GCMs is the calculation of the full three-dimensional character of the climate. The solution of a series of equations (Table 2.1) that describe the movement of energy and momentum and the conservation of mass and water vapour is therefore required. Generally the equations are solved to give the wind field (i.e. mass movement) at the next timestep but models must also include processes such as cloud formation and heat and moisture transport. The first step in obtaining a solution is to specify the atmospheric conditions at a number of 'grid points', obtained by dividing the Earth's surface into a series of rectangles, so that a regular grid results (Figure 2.2). Conditions are specified at each grid point for the surface and several layers in the atmosphere. The equations in Table 2.1 are then solved at each grid point using numerical techniques. Various techniques are available, but all use a timestep approach and an interpolation scheme between grid points.

Table 2.1 Fundamental equations solved in GCMs

1. *Conservation of energy* (the first law of thermodynamics)
   i.e. Input energy = increase in internal energy plus work done
2. *Conservation of momentum* (Newton's second law of motion)
   i.e. Force = mass × acceleration
3. *Conservation of mass* (the continuity equation)
   i.e. The sum of the gradients of the product of density and windspeed in the three orthogonal directions is zero
4. *Ideal gas law* (an approximation to the equation of state)
   i.e. Pressure × volume = gas constant × absolute temperature

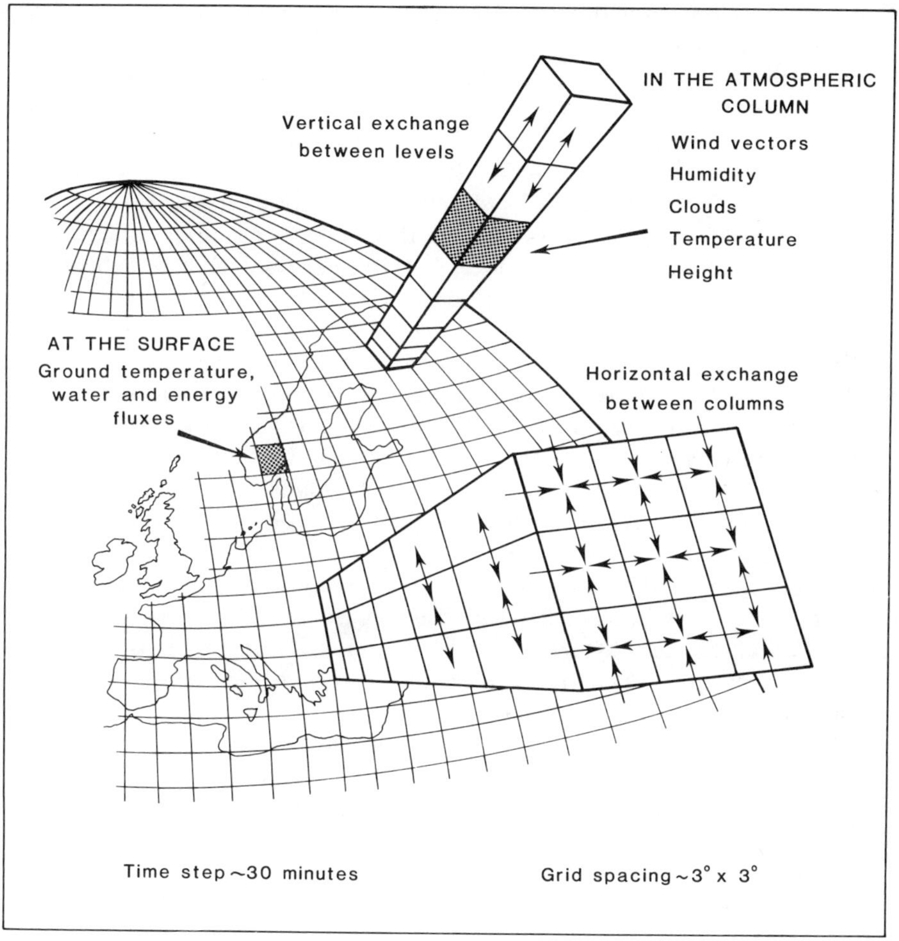

Figure 2.2 Illustration of the basic characteristics of a general circulation climate model. The model is a grid of interacting columns spread across the surface of the globe.

Although GCMs formulated in this way have the potential to approach the real atmospheric situation very closely, at present there are a number of practical and theoretical limitations. The prime practical consideration is one of the time needed for the calculations. To compute each of the basic atmospheric variables at each grid point requires that roughly 10^5 (one hundred thousand) numbers be stored, fetched, recalculated and re-stored at each timestep. This places a strain on the resources of even the largest and fastest computers. Since the accuracy of the model partly depends on the spatial resolution of the grid points and the length of the timestep, a compromise must be made between the resolution desired and the computational facilities available. At present grid points are typically spaced between 3° and 5° of latitude and longitude apart and timesteps of approximately 30 minutes are used. Vertical resolution is obtained by dividing the atmosphere into between six and fifteen levels.

Often the computational time and expense dictate that models be run to simulate a particular season or month rather than a complete annual cycle. Many experiments concerning the effects of increasing atmospheric carbon dioxide on the climate, for example, deal with a 'perpetual January' (or July). As computers improve, this is changing so that full annual cycles are more commonly computed. On the other hand, this increase in computing power (see Figure 1.4) enables us to produce models with finer resolution and more exact solutions of the governing equations, so there still may be a tendency to produce more 'perpetual Januaries'.

It is important to identify the very different aims of those developing and using GCMs as compared to the designers of numerical weather forecast models. The latter are prediction tools while GCMs can represent only probable conditions. For this reason many GCM integrations must be performed and their results averaged before a climate "prediction" can be made.

Computational constraints lead to problems of a more theoretical nature. With a coarse grid spacing small-scale atmospheric motions (termed sub-gridscale), such as thundercloud formation, cannot be modelled, however important they may be in reality for atmospheric dynamics. Fine grid models can only be used for weather prediction because of constraints on computer time, and consequently climate models must rely on some form of parameterization of sub-gridscale processes.

Some of the processes usually incorporated into general circulation climate models are shown in Figure 2.3. Within the atmosphere the modellers adopt an approach similar to that used for the RC models in calculating heating rates (although often computationally simpler), but also often include cloud formation processes as part of the convection and consider in detail the effects of horizontal transports. The interaction between the surface and the near-surface layer of the atmosphere, however, must be parameterized. Detailed consideration of the transfer processes at the surface of the Earth are computationally too demanding for explicit inclusion. Commonly the surface fluxes of momentum, sensible heat and moisture are taken to be proportional to the product of the surface wind speed and the gradient of the property away from the surface. More detailed aspects of GCMs will be considered in Chapter 6.

2.3 HISTORY OF CLIMATE MODELLING

As climate models are often described in terms of a hierarchy, it is often assumed that the simpler models were the first to be developed with the more complex GCMs being developed most recently. This is not the case. The first atmospheric general circulation climate models were being developed in the early 1960s concurrently with the first RC models. On the other hand, energy balance climate models, as they are currently recognized, were not described in the literature until 1969 and the first discussion of two-dimensional SD models was in 1970.

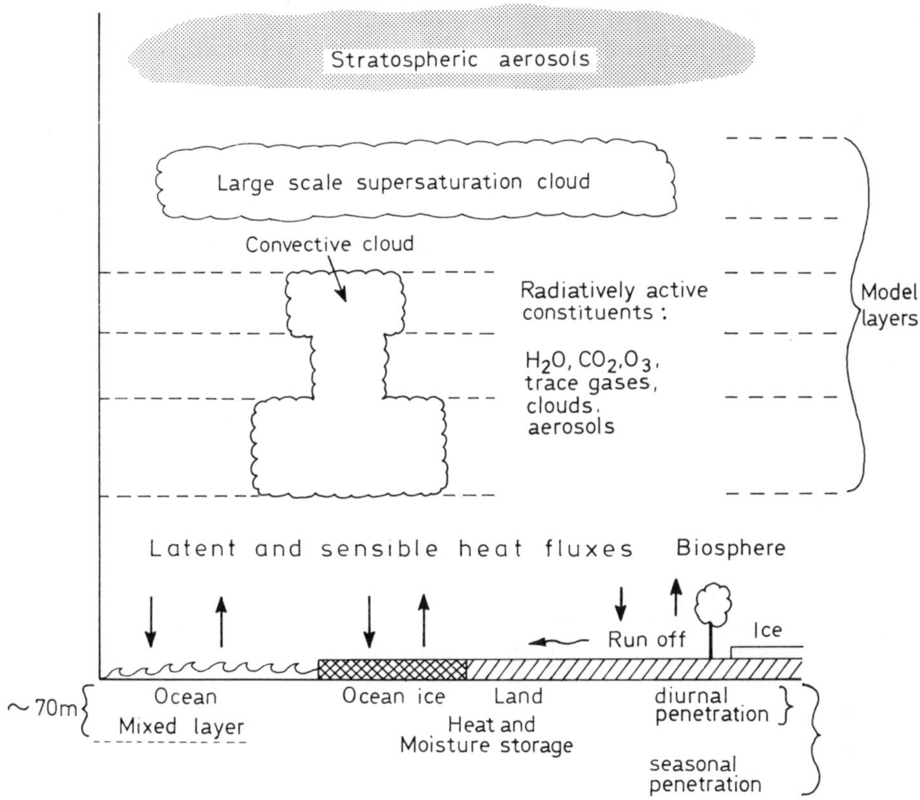

Figure 2.3 Schematic representation of the sub-gridscale structure of a general circulation climate model. In this GCM two types of cloud are treated: gridscale supersaturated clouds and sub-gridscale convective clouds. The surface interaction with the atmosphere is fairly complex and heat and moisture storage within the ground are treated by the use of a two-layer ground model (reproduced by permission of the American Meteorological Society from Hansen *et al.* (1983), *Monthly Weather Review*, **111**, 609–662).

The first atmospheric general circulation climate models were derived directly from numerical models of the atmosphere designed for short-term weather forecasting. These had been developed during the 1950s and around 1960 ideas were being formulated for longer period integrations of these numerical weather prediction schemes. It is in fact rather difficult to identify the transition point in many modelling groups. For example, Syukuro Manabe joined the National Oceanic and Atmospheric Administration's Geophysical Fluid Dynamics Laboratory in the USA in 1959 to collaborate in the numerical weather prediction efforts, but was to go on to become one of the leaders in the climate-modelling community. Scientists concerned with extending numerical prediction schemes to encompass hemispheric or global domains, were also studying the radiative and thermal equilibrium of the Earth–atmosphere system. It was these studies which prompted the design of the RC models, which were once again spearheaded by Manabe; the first of these being published in 1964. Other workers, such as Julian Adem, also expanded the domain of numerical weather prediction schemes in order to derive global climate models. The low-resolution thermodynamic model first described by Adem in 1965 is an interesting type of climate model, since it lies close to the apex of the climate modelling pyramid (Figure 2.1) although the methodology is simpler in nature than that of an atmospheric GCM. Similar in basic composition to an EBM, Adem's model includes, in a highly parameterized way, many dynamic, radiative and surface features and feedback effects, giving it a 'higher' position on the modelling pyramid.

Mikhail Budyko and William Sellers published descriptions of two very similar EBMs within a couple of months of each other in 1969. These models did not depend upon the concepts already established in numerical weather prediction schemes, but attempted to simulate the essentials of the climate system in a simpler way. The EBMs drew upon observational data derived from descriptive climatology, suggesting that major climatic zones are roughly latitudinal. As a consequence of the intrinsically simpler parameterization schemes employed in EBMs, they could be applied to longer timescale changes than the atmospheric GCMs of the time. It was the work by Budyko and Sellers, in which the possibility of alternative stable climatic states for the Earth was identified, which prompted much of the interest in simulation of geological timescale climatic change. Concurrently with these developments, RC models, usually globally averaged, were being applied to questions of atmospheric disturbance including the impact of volcanic eruptions and the possible effects of increasing atmospheric CO_2.

The desire to improve numerical weather forecasting abilities also prompted the fourth type of climate model: the SD model. A primary goal for dynamical climatologists was seen to be the need to account for the observed state of averaged atmospheric motion, temperature and moisture on timescales shorter than seasonal but longer than those characteristic of depression systems. One group of climate modellers preferred to design relatively simple low-resolution SD models to be used to illuminate the nature of the interaction between forced stationary long waves and travelling weather systems. Much of this work was spearheaded in the early 1970s by John Green. Theoretical study of large scale atmospheric eddies and their transfer properties combined with observational work led to the parameterizations employed in two-dimensional climate models.

By 1980, this diverse range of climate models seemed to be in danger of being overshadowed by one type: the atmospheric GCM. Although single-minded individuals persevered with the development of simpler models, considerable funding and almost all the computational power used by climate modellers was being consumed by atmospheric GCMs. Six years later a series of occurrences of apparently correct results being generated for the wrong reason by these highly non-linear and highly complex models has prompted many modelling groups to move backward, hierarchically speaking, in order to try to isolate essential processes responsible for the results which are observed from more comprehensive models. It is clear now that the strategy of intentionally utilizing a hierarchy of models, originally proposed in the 1970s by scientists at the US National Center for Atmospheric Research, is more than sound—it is essential.

A second renaissance in climate modelling during the early 1980s was prompted by the realization that no single feature of the climate system can be studied using a constant framework. Dynamical meteorologists invested considerable effort in attempting to demonstrate that all features of the climate were secondary to motion. Oceanographers drove their models with prescribed surface variables and glaciologists and agriculturalists assumed constant mesoscale inputs. The desire to make climate models more realistic has led to the inclusion of each of these disciplines within the framework of climate modelling and hence to the realization, however unwelcome, that no one discipline can assume constancy in the variables prescribed by the others. Interdisciplinary programmes have led to both rapid growth in insight and near-catastrophic blunders. Also, increasing complexity in narrowly defined areas such as land-surface climatology has forced upon modellers the recognition that other characteristics of their models, such as the diurnal cycle of precipitation, are being poorly predicted.

Joseph Smagorinsky, who pioneered much of the early development in numerical weather prediction and steered the course of one of the flagships of climate modelling—NOAA's Geophysical Fluid Dynamics Laboratory—when commenting on the exponential growth in climate modelling research, noted that at the international conference on numerical weather prediction held in Stockholm in June 1957, which might be considered the first international gathering of climate modellers, the world's expertise comprised about 40 people, while now it would number many thousands.

2.4 SENSITIVITY OF CLIMATE MODELS

Often the development of climate models begins with a series of sensitivity tests. Modellers examine the behaviour of their modelled climate system by altering one component and studying the effect of this change on the model climate.

Equilibrium climatic states

As an example of the change in an internal variable we can consider the variation in the albedo, α, as a function of the mean global temperature. Above a certain temperature, T_g, the planet is ice-free and the value of the albedo is independent of temperature. As it becomes colder we expect the albedo to increase as a direct result of increases in ice and snow cover. Eventually the Earth becomes completely ice covered, at temperature T_i, and further cooling will produce no further albedo change. This could be expressed in the form

$$\left.\begin{aligned} \alpha(T) &= \alpha_i && \text{for } T \leqslant T_i \\ \alpha(T) &= \alpha_g && \text{for } T \geqslant T_g \\ \alpha(T) &= \alpha_g + b(T_g - T) && \text{for } T_i < T < T_g \end{aligned}\right\} \tag{2.2}$$

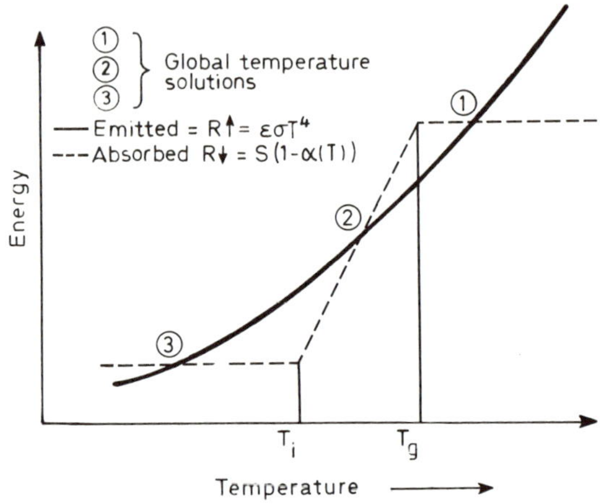

Figure 2.4 The three equilibrium temperature solutions for a zero-dimensional global climate model are shown at the intersection between the curves of emitted infrared radiation $R\uparrow$ and absorbed solar radiation $R\downarrow$. They are: (1) an ice-free Earth; (2) an Earth with some ice; and (3) a completely ice-covered Earth (reproduced by permission of Longman from Henderson-Sellers and Robinson, *Contemporary Climatology*, 1986).

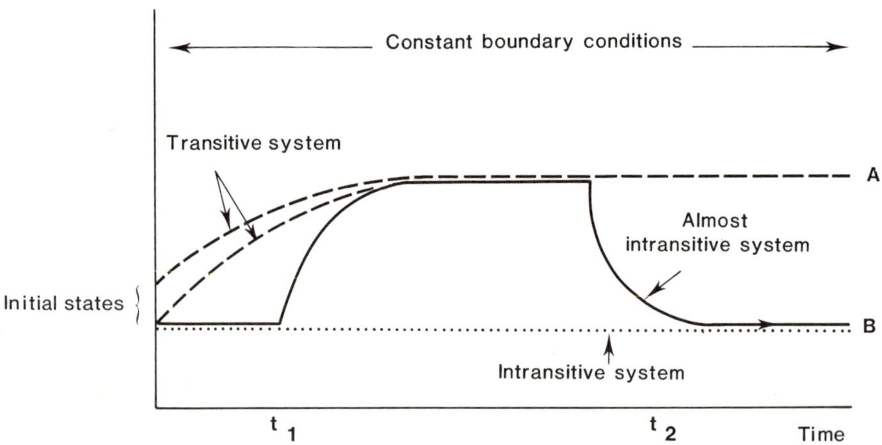

Figure 2.5 The behaviour of the three types of climatic system: transitive, intransitive and almost intransitive with respect to an initial state. In a transitive system two different initial states evolve to the same resultant state, A. An intransitive system exhibits the 'opposite' behaviour with more than one alternative resultant state. The characteristic of an almost intransitive system is that it mimics transitive behaviour for an indeterminate length of time and then 'flips' to an alternative resultant state (reproduced by permission of US National Academy of Sciences, 1975).

T_i is usually assumed to be 273 K but may range between 263 and 283 K. If we are concerned with equilibrium conditions, we can calculate $R\uparrow$ for a series of temperatures and $R\downarrow$ for a series of albedos and show the results graphically. The points of intersection of the curves occur when emitted and absorbed radiation fluxes balance (i.e. $R\uparrow = R\downarrow$) which represent the equilibrium situations (Figure 2.4). Any slight imbalances between the fraction of the incident solar radiation, S, absorbed, $S(1-\alpha(T))$, and the top of the atmosphere emitted longwave flux, approximated by $\epsilon\sigma T^4$ where ϵ is the emissivity, lead to a change in the temperature of the system at the rate $\Delta T/\Delta t$, the changes serving to return the temperature to an equilibrium state. However, there are three equilibrium solutions, as shown in Figure 2.4: a completely glaciated Earth (3), an ice-free Earth (1) and an Earth with some ice (2) (e.g. the present situation of the planet). All are possible.

Equilibrium conditions and transitivity of climate systems

Such a simple model has some very obvious limitations. However, it not only shows some typical lines of approach, it also indicates some of the more general problems associated with the solutions; in particular, the question of whether or not all three equilibrium states identified are 'stable' and capable of persisting for long periods of time.

Many non-linear systems, even ones which are far simpler than the climate system, have a characteristic behaviour termed *almost intransitivity*. This behaviour is illustrated in Figure 2.5. If two different initial states of a system evolve to a single resultant state as time passes the system is termed a *transitive* system. State A for this transitive system would then be considered the solution or normal state and all perturbed situations would be expected to evolve to it. At the other extreme an *intransitive* system has at least two equally acceptable solution states (A and B), depending on the initial state. Difficulty arises when a system exhibits behaviour which mimics transitivity for some time, then flips to the alternative state for another (variable) length of time and then flips back again to the initial state and so on. In such an almost intransitive system it is impossible to determine which is the 'normal state', since either of two states can continue for a long period of time, to be followed by a quite rapid and perhaps unpredictable change to the other.

At present geological and historical data are not detailed enough to determine for certain which of these system types is typical of the Earth's climate. In the case of the Earth the alternative climate need not be so catastrophic as a complete glaciation. It is easy to see that should the climate turn out to be almost intransitive successful climate modelling will be extremely difficult.

Stability of model results

Great care must be taken in choosing the constants for any parameterization scheme in any model. If values have been determined solely from empirical

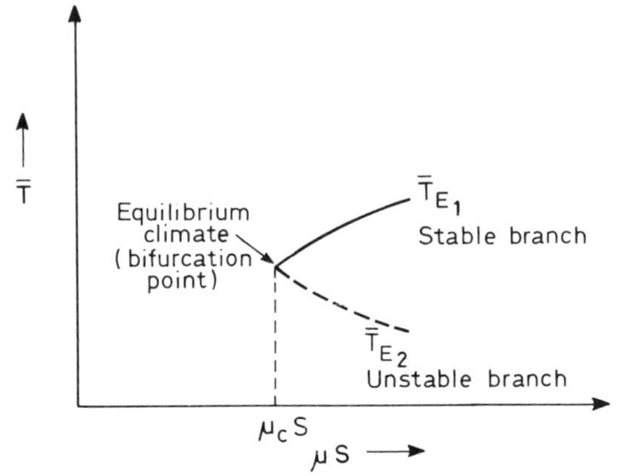

Figure 2.6 The equilibrium climate bifurcation point. For values of the solar luminosity given by μS where μ is a fractional premultiplier of the solar constant S, such that $\mu S > \mu_c S$ there are two solutions, whereas below this critical value no solutions exist. Changes in solar radiation lead to either a stable or unstable equilibrium climate illustrated here by the two equilibrium branches.

evidence it may be that they are appropriate only for the present day, with the result that the model is likely to be constrained to predict the present-day situation and the less likely it is to be able to respond realistically to perturbations.

For 'external stability' we can test the response of the model to perturbations in the solar constant since this is a convenient method of exploring climate model structure. Figure 2.6 shows the way in which $\bar{T}$ changes as the total incident radiation, μS, changes. Reduction of the solar constant to some critical value ($\mu_c S$) means that the number of solutions is reduced from two to one. Below $\mu_c S$ no solution is possible. This point is termed the bifurcation point. For values of incoming radiation, μS, less than $\mu_c S$ temperatures are so low that the albedo, $\alpha(T,\phi)$, becomes very close to or equal to 1 and thus it is impossible to regain energy balance and the global temperature, $\bar{T}$, approaches minus infinity. However, if some limit is put on how high the albedo may become as is usually the case, e.g. $\alpha \leqslant 0.75$, the solution becomes what might be described as an ice-covered Earth.

'Internal stability' concerns the response of each branch in Figure 2.6 to perturbations from equilibrium which are created by internal factors. To determine if temperatures will return to equilibrium after the perturbation, we can use a time-dependent formulation and postulate a new value for $\bar{T}$ which is close to the equilibrium one already calculated at that level of μS. This change can be computed iteratively until it is determined whether the values do regain the original $\bar{T}$ solution. If it is regained, then the solution is said to be 'internally stable'. In the case shown in Figure 2.6 only the top branch is stable because the model preserves $\bar{T}$ as proportional to μS. Using this method it is possible to determine whether the model is transitive or intransitive. The identification of almost intransitivity is not possible in this manner.

2.5 PARAMETERIZATION OF CLIMATIC PROCESSES

The climate system is a physico-chemico-biological system possessing infinite degrees of freedom. Any attempt to model such a highly complex system is fraught with dangers. It is (unfortunately) necessary to represent a distinct part, or more usually many distinct parts, of the complete system by inaccurate or semi-empirical mathematical expressions. Worse still is the need to neglect completely many parts of the complete and highly complex system. This process of neglect/semi-empirical or inaccurate representation is termed parameterization. Parameterization can take many forms. The simplest form is the null parameterization where a process, or more properly a group of processes, is ignored. The decision to neglect these can only be made after a detailed consideration of their importance relative to other processes being modelled. Unnecessary computing time should not be spent on processes that can be adequately represented in some simpler way, or on processes that have relatively little effect on the climate at these scales.

Climatological specification is a form of parameterization which is widely used in most types of model. In GCMs it is not uncommon to specify oceanic temperatures (with a seasonal variation) and in some of these models the cloud field is specified. This is an unwise procedure for climate sensitivity experiments, since important feedback features of the climate system have been suppressed. Only slightly less hazardous is the procedure by which processes are parameterized by relating them with reference to present-day observations: the constants or functions describing the relationship between variables are 'tuned' to obtain agreement. It is important that physically unrelated processes are not tuned together by this method. For example, the association of gradients in two different variables need not mean that the two are physically related. At best, this procedure presumes that constants and relationships appropriate to today's climate will still be applicable should some aspect of the climate alter. Seasonal variation can provide a greater range of conditions than various climate-changing mechanisms are thought likely to provide. Hence parameterizations tuned to annual data sets can, and should, be tested in a seasonal mode.

The most advanced parameterizations have a theoretical justification. For instance, in some two-dimensional zonally averaged dynamical models the fluxes of heat and momentum are parameterized via baroclinic theory (in which the eddy fluxes are related to the latitudinal temperature gradient). The parameterization of radiative transfer in clear skies is another example. All that needs to be known is the vertical variation of temperature and humidity. Unfortunately, these parameterizations can lead to problems of uneven weighting because another process of equal importance cannot be adequately treated. In the case of heat and momentum transport by eddies, the contribution to these fluxes from stationary waves forced primarily by the orography and the land/ocean thermal contrast cannot be so easily considered. In radiation schemes, the parameterization of cloudy sky processes is not as advanced as for clear skies.

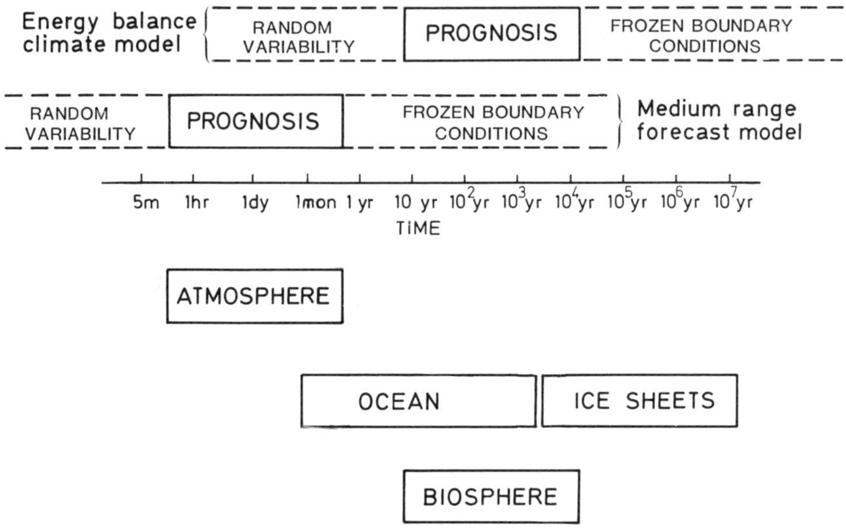

Figure 2.7 The importance of different temporal scales changes as a function of the type of model. The domain in which the model simulates the behaviour of the system is labelled 'prognosis'. It is expected that processes which fluctuate very rapidly compared with the prognostic timescales will contribute only small random variability to the model predictions, while processes which fluctuate very slowly compared with the prognostic timescale can be assumed to be constant. Two types of model are shown: an EBM and a medium-range weather forecast model.

As the climate system depends upon scales of motion and interactions ranging from molecular to planetary dimensions, and from timescales of nanoseconds to geological eras, parameterizations are a necessary part of the modelling process. In particular a decision is generally made very early in model construction about the range of space- and timescales which will be modelled explicitly. Figure 2.7 illustrates the difficulty faced by all modellers. The constraints of computer time and costs and data availability restrict the prognostic (or predictive) mode. Outside this range there are 'frozen' boundary conditions and 'random variability'. Thus the two examples shown in Figure 2.7 illustrate the range of prognostic computations for (i) a medium-range weather forecast model, and (ii) an EBM focused upon examining the effect of Milankovitch variations on the climate. In both cases longer timescales than those of concern to the modeller are considered as invariant, and shorter timescales are neglected as being random statistical fluctuations, the details of which are of too short a period to be of interest.

Parameterizations must be mutually consistent. For instance, if two processes produce feedback effects of opposite sign it is important that one process is not considered in the other's absence. An example is the effect that clouds have on the radiative heating of the atmosphere. Longwave radiation causes a comparatively rapid cooling at the cloud top, whereas the absorption of solar radiation results in heating. To consider the effect of clouds on only

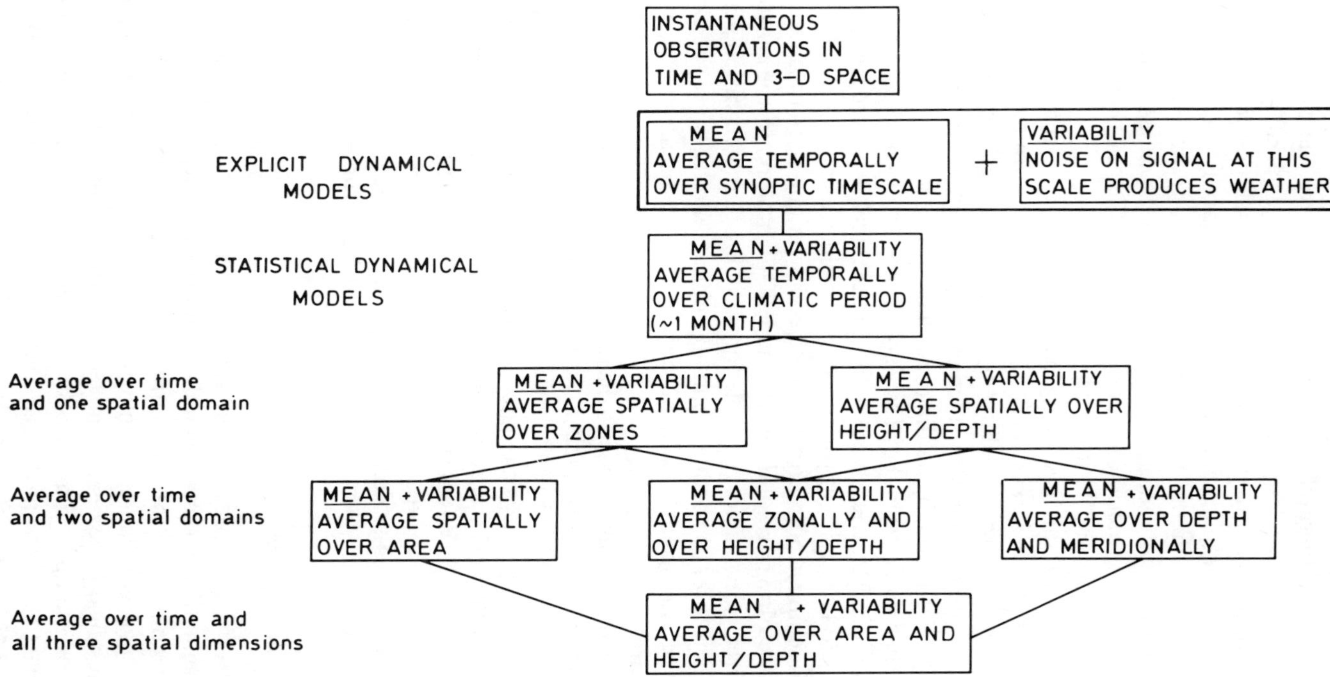

Figure 2.8 A hierarchical scheme for the averaging of climatological variables. In the lower half of the figure the representations of the climate system on the right-hand side involve averaging first over the atmospheric column, whereas the representations on the left-hand side involve zonal averaging first.

one of the two radiation fields may be worse than neglecting the effect of clouds entirely.

Figure 2.8 portrays a hierarchical averaging scheme for the climate system. The averaging processes are described in terms of a single variable, which could be as simple a component of the climate system as temperature, but could alternatively be, for example, representative of the carbon budget. There are two averaging subsystems in the lower part of the diagram, the one on the right-hand side being based on an initial averaging of the mean state in the vertical, followed by zonal and/or meridional averaging, while the one on the left-hand side is averaged first around latitude zones.

A traditional view of this averaging diagram would be that the simplest approximations to the climate system (models) lie at the bottom of the diagram (cf. the base of the climate-modelling pyramid: Figure 2.1) with increasing resolution being synonymous with increasing (and perhaps more desirable) complexity on ascent through the diagram. The apex of this diagram would be presumably that radiative and diffusive processes would be described at the molecular level in GCMs. Clearly such an ultimate goal is absurd, although it sometimes seems to be consistent with the desire for increasing complexity within a few GCM modelling groups. An alternative view might be that some of the more sophisticated lower-resolution SD models might contain the maximum information currently available/verifiable for very long-term integration periods. These would, therefore, be adequate and appropriate models since the climate system over long timescales would be deemed to be insensitive to higher-resolution features.

Thus the key element in any model is the method of parameterization, whereby processes that cannot be treated explicitly are instead related to variables that are considered directly in the model. An example is in EBMs where only the surface temperature may be explicitly calculated. Since poleward transport of heat by atmospheric motions is important, this transport has to be parameterized in some way relating to the surface temperature, such as the latitudinal temperature gradient. In GCMs, those processes which operate on scales too small to be resolved by the model (sub-gridscale processes), like convective clouds, can, and do, exert influence on the atmosphere and must be parameterized.

Interactions in the climate system

The interactions between processes in any model of the climate are crucially important. 'Spaghetti' diagrams which show all these interactions are often used to illustrate the complexity of incorporating them all adequately. Figure 2.9 shows one such spaghetti diagram from which the names of the processes (boxes) and variables (circles) have been removed. For the present-day climate being simulated by, for example an atmospheric GCM, the top circle might contain the word TEMPERATURE, the next three boxes SENSIBLE HEAT FLUX, RADIATION and EVAPORATION and so on down to the lowest circle which might contain the words SOIL MOISTURE. It is an interesting exercise to see how many different

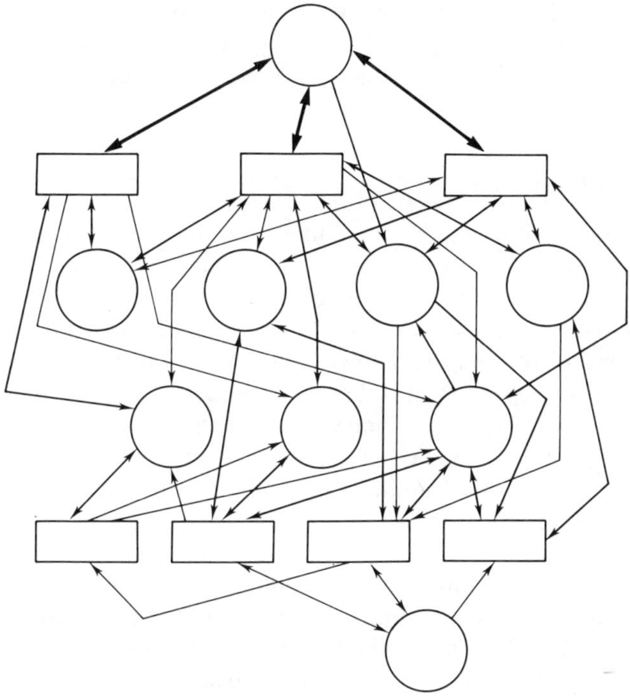

Figure 2.9 A generalized 'spaghetti' diagram for climate models. The arrows show the links between a set of unnamed processes and variables. The processes which could be entered in the boxes and the variables in the circles depend crucially on the timescale being modelled and the type of model employed.

ways the boxes and circles can be filled. The demarcation between circles and boxes can be questioned according to the circumstances in mind, as can the relative thicknesses of the connecting arrows.

All the same, thinking up processes that fit is more challenging than drawing the arrows since most things in the climate system are linked. Another interesting exercise is to think how the processes named would change if, for example, the climate model was being designed to simulate glacial cycles. The model type in this case might be an EBM and this time the top circle might again contain the word TEMPERATURE, but the next three might denote ABSORBED SOLAR RADIATION, EMITTED LONGWAVE RADIATION and MERIDIONAL TRANSFER and so on down to the lowest circle which could relate to CLOUD AMOUNT. These two possible sets of processes that fit into Figure 2.9 emphasize a most important concept in climate modelling: the relative importance of processes and the way different processes interlink is a strong function of the timescale being modelled. The whole concept of parameterization is subsumed by this assertion. Establishing whether a system is likely to be sensitive to the parameterization used for a particular process often depends upon the response time of that feature as compared with other 'interactive' features. It is pointless to invoke a highly complex, or exceedingly simplistic, parameterization if it has been constructed for a timescale different from that of the other processes and linkages in the model. The adage 'choosing horses for courses' is fundamental to the art of climate modelling.

The need for observations

The oceans play a vital role in the climate system. A reasonable estimate of the thermal response time of the oceanic mixed layer is around 7–8 years. This is important for two reasons. It is unlikely that any perturbation which operates on timescales significantly shorter than 7–8 years can cause a significant climatic response. Such short-period fluctuations would be damped by the thermal inertia of the oceans. The second point is that model simulations need to cover at least this time period (usually much longer) so that the mixed layer of the oceans is able to attain equilibrium. It seems likely from these calculations that the necessary timescale of model simulations needs to be increased to $\sim$10–20 years when they include some form of interactive ocean.

The interaction between the ocean and the atmosphere can be considered in several ways. Rather simplistic, but commonly accepted 5–10 years ago, was the 'swamp' model in which the ocean acts only as an unlimited source of moisture. Many GCMs still use fixed ocean temperatures based on observed average values. Naturally, it is very difficult in such a model to disturb the climate away from present-day conditions when such large areas of the globe remain unchanged. At present, computation of the heat storage of the mixed layer of the ocean is the most common approach. Usually the maximum depth of the mixed layer is prescribed to be around 70–100 m, and the lower deep ocean layer acts only as an infinite source and sink for water. This assumption is perhaps less appropriate for timescales $\geqslant$100 years when the transfer of heat to lower levels becomes significant.

Horizontal and vertical motions play a vital role in the global ocean. Although the processes in the atmosphere and the ocean are similar, the temperature changes and some mixing and transport in the oceans are very much slower than in the atmosphere. Thus even if satisfactory oceanic circulation models are developed there remains the problem of linking an atmospheric model with a timestep of around 20–30 minutes (or less) to an oceanic model that needs between 20 and 30 years to come to equilibrium. The difficulty of linking models of these two systems is discussed in Chapter 6.

Many models need observed values for part of their input especially in order to specify the boundary conditions and all require observational data with which to compare their results. Some variables, such as surface pressure, are available worldwide and pose only the problem of evaluating the accuracy of the observed data set. Others, however, are sparse in either time or space. Knowledge of sea-ice extent is largely dependent on satellite observations, so that there is only a short observational record. Thus it is difficult to compare observations with any long-term average values obtained from models. Similarly, global coverage for cloud observations is relatively recent, so that it is difficult even to initialize models with the present-day cloud distribution. Certainly, the improved cloud climatologies which are being developed as more satellite observations become available are needed before more complex models can be validated.

2.6 SIMULATION OF THE FULL, INTERACTING CLIMATE SYSTEM: ONE GOAL OF MODELLING

Despite their limitations, GCMs represent the most complete type of climate model currently available. They illustrate the tremendous advances in our understanding of the atmosphere and our ability to model it over the 25 years since the first numerical climate models were produced. They do not, however, incorporate all aspects of the climate system and are therefore not at the absolute apex of the pyramid in Figure 2.1. They can provide a great deal of information about the present climate and the possible effects of future perturbations. That these predictions are often contradictory is inevitable, given our relatively poor knowledge of present conditions and understanding of the controlling processes. If a model is developed on sound physical principles, incorporates rational parameterization schemes, accounts for the major processes acting in the atmosphere and has been adequately tested against the present conditions, its results should be treated with respect. The results provide at least an indication of the possible future climate conditions created by a perturbation in the forces which control our present climate.

The rest of the book is structured so that the concepts upon which full three-dimensional models are based are introduced sequentially. Chapter 3 underlines the fundamental basis of climate modelling: the energy balance. Chapter 4 describes radiation in the context of RC models and Chapter 5 introduces the importance of meridional transport particularly of mass and energy. The overt goal of the text is therefore clear: we are aiming towards Chapter 6 in which GCMs, including coupled atmosphere–ocean models, are described. The other equally valid and important goal is less obvious. Throughout the book we have tried to choose examples to illustrate and enhance *understanding of the mechanisms* controlling the climate, their complexities, time- and space-scales and interactions. Both goals are worthy of considerable effort.

RECOMMENDED READING

Adem, J. (1965) Experiments aiming at monthly and seasonal numerical weather prediction, *Mon. Wea. Rev.*, **93**, 495–503.

Adem, J. (1979) Low resolution thermodynamic grid models, *Dyn. Atmos. Ocean.*, **3**, 433–451.

Bourke, W., McAveney, B., Puri, K., and Thurling, R. (1977) Global modelling of atmospheric flow by spectral methods. In J. Chang (ed.) *Methods in Computational Physics*, Vol. 17, Academic Press, New York, pp. 267–324.

Budyko, M. I. (1969) The effect of solar radiation variations on the climate of the Earth, *Tellus*, **21**, 611–619.

Chang, J. (ed.) (1977) General circulation models of the atmosphere. *Methods in Computational Physics*, Vol. 17, Academic Press, New York, 337pp.

Gates, W. L. (1979) The effect of the ocean on the atmospheric general circulation, *Dyn. Atmos. Ocean.*, **3**, 95–109.

Green, J. S. A. (1970) Transfer properties of the large scale eddies and the general circulation of the atmosphere, *Quart. J. Roy. Meteor. Soc.*, **96**, 157–185.

Hansen, J. E., Johnson, D., Lacis, A. A., Lebedeff, S., Lee, P., Rind, D., and Russell, G. (1981) Climate impact of increasing atmospheric CO_2, *Science*, **213**, 957–1001.

Hansen, J. E., Russell, G., Rind, D., Stone, P. H., Lacis, A. A., and Lebedeff, S. (1983) Efficient three dimensional models for climate studies. Models I and II, *Mon. Wea. Rev.*, **111**, 609–662.

Hasselmann, K. (1976) Stochastic climate models, Part 1. Theory, *Tellus*, **28**, 473–485.

Held, I. M., and Suarez, M. J. (1978) A two level primitive equation model designed for climate sensitivity experiments, *J. Atmos. Sci.*, **35**, 206–229.

Manabe, S., and Bryan, K. (1969) Climate calculations with a combined ocean atmosphere model, *J. Atmos. Sci.*, **26**, 786–789.

Manabe, S., and Möller, F. (1961) On the radiative equilibrium and heat balance of the atmosphere, *Mon. Wea. Rev.*, **89**, 503–532.

Manabe, S., and Strickler, R. F. (1964) Thermal equilibrium of the atmosphere with a convective adjustment, *J. Atmos. Sci.*, **21**, 361–385.

Potter, G. L., Ellsaesser, H. W., MacCracken, M. C., and Mitchell, C. S. (1981) Climate change and cloud feedback: the possible radiative effects of latitudinal redistribution, *J. Atmos. Sci.*, **38**, 489–493.

Saltzman, B. (1978) A survey of statistical dynamical models of terrestrial climate, *Advances in Geophysics*, **20**, 183–304.

Shine, K. P., and Henderson-Sellers, A. (1983) Modelling climate and the nature of climate models: a review, *J. Climatol.*, **3**, 81–94.

Smagorinsky, J. (1983) The beginnings of numerical weather prediction and general circulation modeling: early recollections. In B. Saltzman (ed.) *Theory of Climate*, Academic Press, New York, 3–38.

Stone, P. H. (1973) The effects of large scale eddies on climatic change, *J. Atmos. Sci.*, **30**, 521–529.

Thompson, S. L., and Schneider, S. H. (1979) A seasonal zonal energy balance climate model with an interactive lower layer, *J. Geophys. Res.*, **84**, 2401–2414.

Washington, W. M., Semtner, A. J. jr, Meehl, G. A., Knight, D. J., and Meyer, T. A. (1980) A general circulation experiment with a coupled atmosphere, ocean and sea ice model, *J. Phys. Oceanogr.*, **10**, 1887–1908.

Wiin-Nielsen, A. C., and Sela, J. (1971) On the transport of quasi-geostrophic potential vorticity, *Mon. Wea. Rev.*, **99**, 447–459.

Chapter 3

Energy Balance Models

'The more it snows,
(Tiddely pom),
The more it goes,
(Tiddely pom),
The more it goes,
(Tiddely pom),
On snowing . . .'

From *The House at Pooh Corner*, by A. A. Milne (1928). Reproduced by permission of Methuen Children's Books, McClelland and Stewart, Toronto and E. P. Dutton, a division of NAL Penguin Inc.

3.1 BALANCING THE PLANETARY RADIATION BUDGET

Balancing the planetary radiation budget offers a first approximation to a model of the Earth's climate. The radiation streams and the equator-to-pole energy transport are the fundamental processes of the climate system. It is these processes which are incorporated in EBMs. Models of this type had been studied for a number of years but renewed interest was stimulated by the independent results of Budyko and of Sellers in 1969.

Many of the questions which these studies raised have since been answered, but these models remain interesting tools for studying climate. This chapter describes how EBMs are constructed and outlines how the models have been used both to study and to illustrate components of the climate system.

3.2 THE STRUCTURE OF ENERGY BALANCE MODELS

The simplest method of considering the climate system of the Earth, and indeed of any planet, is in terms of its global energy balance. Viewing the Earth from outside, one observes an amount of radiation input which is balanced (in the long term) by an amount of radiation output. Since over 70 per cent of the energy which drives the climate system is first absorbed at the surface, the surface albedo will be predominant in controlling energy input to the climate system. The output of energy will be controlled by the temperature of the Earth but also by the transparency of the atmosphere to this outgoing thermal radiation. An EBM can take two very simple forms. The first form, the zero-dimensional model,

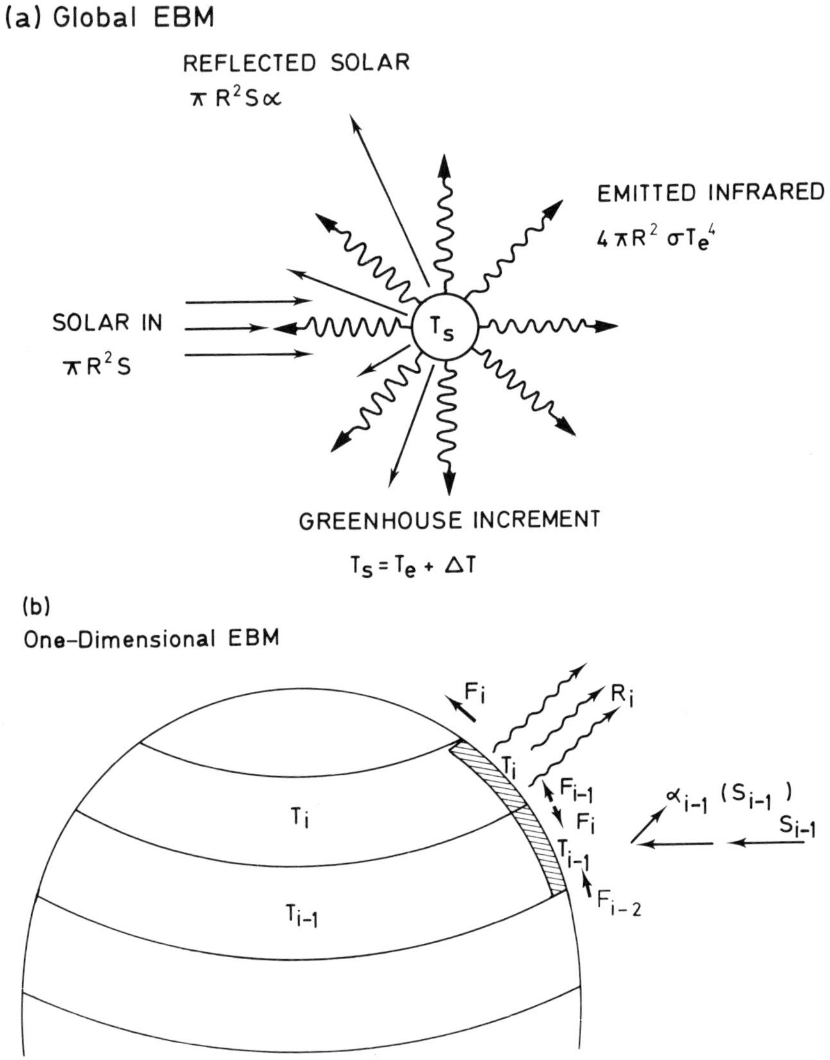

Figure 3.1 Energy transfers in (a) a global EBM; and (b) a full zonal EBM.

considers the Earth as a single point in spacc having a global mean effective temperature T_e. The second form of the EBM considers the temperature as being latitudinally resolved. Figure 3.1 illustrates these two approaches. In the first case the climate can be simulated by considering the radiation balance. The total energy received from the Sun per unit time is $\pi R^2 S$ where R is the radius of the Earth. The total area of the Earth is, however, $4\pi R^2$. Therefore the time-averaged energy input rate is $S/4$ over the whole Earth. Hence,

$$(1-\alpha)S/4 = \sigma\, T_e^4 \tag{3.1}$$

where α is the planetary or system albedo, S is the solar constant (1370 W m^{-2}) and σ is the Stefan–Boltzmann constant. If the atmosphere of the planet contains gases which absorb thermal radiation, then the surface temperature, T_s, will be greater than the effective temperature, T_e. The increment ΔT is known as the *greenhouse increment* and depends upon the efficiency of the infrared absorption. Thus the surface temperature can be calculated if ΔT is known since

$$T_s = T_e + \Delta T \tag{3.2}$$

For the Earth the greenhouse increment due to the present atmosphere is about $\Delta T = 33$ K and hence combining equations (3.1) and (3.2) gives, for $\alpha = 0.3$, $T_s = 288$ K. (Note that the only prognostic variable in an EBM is the temperature, characterized as a surface temperature.)

If the planetary features were different, for example if the solar luminosity were $S = 2619$ W m^{-2} and $\alpha = 0.7$, then $T_e = 242$ K. These are the values appropriate to the planet Venus which, while being closer to the Sun and hence enjoying greater incident solar radiation, is almost completely cloud covered and thus has a very high planetary albedo. The albedo dominates the radiation balance resulting in an effective temperature which is slightly lower than that of the Earth. However, the atmosphere of Venus is extremely dense and composed almost entirely of carbon dioxide. Hence there is a very much greater greenhouse effect on Venus. The surface temperature of Venus has been found by spacecraft to be ~ 730 K and, although it is now believed that the whole of this ΔT is not due to greeenhouse absorbers, they certainly contribute substantially. Moreover, it is believed that there is a further surface heating caused by adiabatic heating in large regions of descending air.

In a simple EBM the incoming and outgoing energy for the globe are balanced and a single climatic variable (the surface temperature, T_s) is calculated, i.e. T is the dependent variable for which the 'climate equations' are solved. The rate of change of temperature, T, with time, t, is caused by a difference between the top-of-the-atmosphere (or planetary) net incoming, $R\downarrow$, and net outgoing, $R\uparrow$, radiative fluxes (per unit area).

$$mc\Delta T/\Delta t = (R\downarrow - R\uparrow)A_e \tag{3.3}$$

where A_e is the area of the Earth, c is the specific heat capacity of the system and m is the mass of the system. This is a very general equation with a variety of uses. If, for example, the system we wish to model is an outdoor swimming pool we can calculate the rate of temperature change in timesteps of 1 day from equation (3.3). Suppose the pool has surface dimensions 30×10 m, is well mixed and is 2 m deep. Since 4200 J of energy are needed to raise the temperature of 1 kg of water by 1 K (4200 J kg^{-1} K^{-1} is the specific heat of water), and 1 m^3 of water has a mass of 1000 kg, the pool has a total heat capacity equal to 2.52×10^9 J K^{-1}. If we assume that the difference between the absorbed radiation and the emitted radiation from the pool ($R\downarrow - R\uparrow$) is 20 W m^{-2} for

24 hours, then the difference in energy content of the pool for each 24-hour timestep is $20\times30\times10\times24\times60\times60$ J. Then, from equation (3.3)

$$2.52\times10^{9}\ \Delta T = 20\times30\times10\times24\times60\times60$$

$$\Delta T \text{ (in 1 day)} = \frac{5.184\times10^{8}}{2.52\times10^{9}} \simeq 0.2\ \mathrm{K} \tag{3.4}$$

Thus at this rate it would take about a month to raise the temperature of the pool water by 6 K.

On the Earth the value of c is largely determined by the oceans. The specific heat (J kg^{-1} K^{-1}) for water is around four times that for air and the mass of the ocean is also much greater than that of the atmosphere. For instance, if we assume that the energy is absorbed in the first 70 m of the ocean (the average global depth of the top or mixed layer) and that approximately 70 per cent of the Earth's surface is covered by oceans, then the value for C (the total heat capacity) comes from

$$C = \varrho_w c_w d A_E\ 0.7 = 1.05\times10^{23}\ \mathrm{J\ m^{-2}\ K^{-1}} \tag{3.5}$$

where ϱ_w is the density of water, c_w its specific heat capacity and d is the depth of the mixed layer and A_E is the Earth's surface area.

For our simple EBM of the Earth, the energy emitted, $R\uparrow$, can be estimated using the Stefan–Boltzmann law and the surface temperature, T. This value must be corrected to take into account the infrared transmissivity of the atmosphere τ_a since $R\uparrow$ is the planetary flux. Therefore we can write

$$R\uparrow \simeq \epsilon\sigma T^4 \tau_a \tag{3.6}$$

The absorbed energy, $R\downarrow$, is a function of the solar flux, S, and the planetary albedo such that $R\downarrow = (1-\alpha)S/4$.

Equation (3.3) therefore becomes

$$\frac{\Delta T}{\Delta t} = \frac{1}{C}\left\{\frac{S}{4}\ (1-\alpha) - \epsilon\tau_a\sigma T^4\right\} \tag{3.7}$$

This equation can be used to ascertain the equilibrium climatic state by setting $\Delta T/\Delta t = 0$. This use is complementary to the timestep mode described above. The result represents an 'ultimate' or equilibrium solution of the equation when the change in temperature has ceased. In this case

$$(1-\alpha)S/4 = \epsilon\tau_a\sigma T^4 \tag{3.8}$$

Using values of $S = 1370$ W m^{-2}, $\alpha = 0.3$, $\epsilon\tau_a = 0.62$ and $\sigma = 5.67\times10^{-8}$ W m^{-2} K^{-4} gives rise to a surface temperature of 287 K which is in good agreement with the globally averaged surface temperature today.

An alternative use of equation (3.7) is similar to the calculation of the swimming-pool warming rate made above. Here a timestep calculation of the change in T is made. This could be a response to an 'external' forcing agent, such as a change in solar flux or in the heat capacity of the oceans resulting from changes in their depth or area. Alternatively, the response could be determined by an 'interactive' climate calculation when one of the internal variables (e.g. α) alters.

In the case where we consider each latitude zone independently

$$S_i(1-\alpha(T_i)) = R\uparrow(T_i) + F(T_i) \tag{3.9}$$

Note that we now have an additional term $F(T_i)$ which refers to the loss of energy by a latitude zone i to its colder neighbour or neighbours. So far we have ignored any storage by the system since we have been considering the climate on timescales where the net loss or gain of stored energy is small. Any stored energy would simply appear as an additional term, $Q(T_i)$ on the right-hand side of equation (3.9).

Since the zero-dimensional model (equation (3.8)) is a simplification of equation (3.9), further discussion will consider the latitudinally resolved model and look in detail at the role of the terms involved.

Each of the terms in equation (3.9) is a function of the predicted variable T_i. The surface albedo is influenced by temperature in that it is increased drastically when ice and snow are able to form. The radiation emitted to space is proportional to T^4, although over the temperature range of interest ($\sim$250–300 K) this dependence can be considered linear. The horizontal flux out of the zone is a function of the difference between the zonal temperature and the global mean temperature. The storage term, an attempt to account for the effect of the oceans, depends on the difference between the present temperature and a long-term mean temperature. The albedo is described by a simple step function such that

$$\alpha_i \equiv \alpha(T_i) \begin{cases} = 0.6 & T_i \leqslant T_c \\ = 0.3 & T_i > T_c \end{cases} \tag{3.10}$$

which represents the albedo increasing at the snowline; T_c, the temperature at the snowline, is typically between -10 and 0 °C. Radiation leaving the top of the latitude zone can be approximated by

$$R_i \equiv R\uparrow(T_i) = A + BT_i \tag{3.11}$$

where A and B are empirically determined constants designed to account for the greenhouse effect of clouds, water vapour and CO_2. The rate of transport of energy can be represented as being proportional to the difference between the zonal temperature and the global mean temperature by

$$F_i \equiv F(T_i) = K(T_i - \bar{T}) \tag{3.12}$$

where K is an empirical constant.

Incorporation of equations (3.11) and (3.12) into equation (3.9) forms an equation which can be rearranged to give the equation

$$T_i = \frac{S_i(1-\alpha_i) + K\bar{T} - A}{B+K} \tag{3.13}$$

Given a first guess temperature distribution, successive applications of this equation will eventually yield an equilibrium solution. An alternative course of action is to calculate the time evolution of the model climate explicitly by including a term representing the thermal capacity of the system. The former method results in computationally faster results but the latter allows for more experimentation.

Such models are relatively simple to construct on a microcomputer (see Section 3.4).

3.3 PARAMETERIZING THE CLIMATE SYSTEM FOR ENERGY BALANCE MODELS

The model described above illustrates the basic principles of energy balance climate modelling. In this section we shall consider further each of the parameterization schemes and how they are developed.

As mentioned previously, the first EBMs were found to be alarmingly sensitive to changes in the solar constant. Small reductions in solar constant appeared to cause catastrophic and irreversible glaciations of the entire planet. Such an effect, although extreme, suggests that such models might be utilized in studying large scale glaciation cycles. This is indeed the case, but some preparation and background work on the mechanisms in the model need to be undertaken before glaciation cycles can be simulated.

Albedo

The albedo parameterization in EBMs is based simply on the surface albedo being greater when the temperature is low enough to allow snow and ice formation. Two simple parameterizations are that the albedo increases instantaneously to an ice-covered value (equation (3.10)), and a description, which might seem more appropriate, that the albedo increases linearly over a temperature interval within which the Earth can be said to be becoming increasingly snow-covered.

$$\begin{aligned} \alpha(T_i) &= b(\phi) - 0.009T_i \qquad && T_i < 283\ \text{K} \\ \alpha(T_i) &= b(\phi) - 0.009 \times 283 \qquad && T_i \geqslant 283\ \text{K} \end{aligned} \tag{3.14}$$

Using empirical constants, $b(\phi)$, allows for the inclusion of a latitudinal variation of ice-free albedo which is not affected by temperature. The change in planetary albedo at the poles is then around half of that at the equator when the ice-free surface is replaced with an ice-covered one. The sensitivity is reduced by a factor of two but remains too high to explain the faint Sun–warm early Earth paradox. This conundrum stems from the inference that although the solar luminosity was only about 70 per cent of its present value, the surface of the Earth seems not to have been glaciated to the extent which would be suggested by these EBM calculations during the first aeon of its history (i.e. during the period from 3.5 to 4.5 thousand million years ago there is little evidence for widespread glaciation).

The explanation for the apparent gross instability of the Earth–atmosphere system to small perturbations in solar constant lies in the close coupling in these models of the temperature and planetary albedo. This is generally not borne out in the climate system since, although the surface albedo is strongly influenced by temperature, the planetary albedo is affected by the presence of clouds and is also a function of latitude. For example, as latitude increases, the effect on the planetary albedo of adding more snow or ice tends to decrease.

The fundamental flaw in albedo parameterization is the assumption of a very strong connection between the planetary albedo and the surface albedo. Clouds are responsible for the reflection of 70–80 per cent of the radiation which is reflected by the Earth. There is no clear relationship between surface temperature and cloudiness which further reduces the connection between surface temperature and planetary albedo. It would appear at first glance that clouds have been ignored in the formulation of EBMs as described in the preceding two sections. The effect of an increase in cloudiness on the amount of absorbed solar radiation is almost exactly countered by the effect of clouds in retaining a greater proportion of emitted infrared radiation.

Outgoing infrared radiation

The Earth is constantly emitting radiation. Some of the radiation which is emitted is absorbed by the atmosphere and re-emitted back to the ground. Parameterizations will involve some method of accounting for this greenhouse effect. One formulation is to match outgoing longwave radiation to surface temperature and to devise a linear relationship between the two. This was the method which was included in equation (3.11). An alternative formulation is to modify the black body flux by some factor which accounts for the amount re-emitted downwards by the atmosphere, e.g.

$$R_i = \sigma T_i^4 [1 - m_i \tanh(19 T_i^6 \times 10^{-16})] \tag{3.15}$$

where m_i is the factor representing the atmospheric opacity. This formulation was derived empirically by Sellers. All parameterizations of infrared radiation follow these basic lines in EBMs.

Heat transport

The simplest form of heat transport which may be incorporated into an EBM is that of equation (3.12). Here the flux out of a latitude zone is equal to some constant multiplied by the difference between the average temperature of the zone and the global mean temperature. A more complex method is to consider each of the transporting mechanisms separately, with the flux divergence being given by

$$\mathrm{div}(F) = \frac{1}{\cos\phi}\frac{\partial}{\partial y}\left[\cos\phi(F_o + F_A + F_q)\right] \tag{3.16}$$

the three terms on the right representing transports due to ocean, atmosphere and latent heat:

$$\begin{aligned} F_o &= -K_o \frac{\partial T}{\partial y} \\ F_a &= -K_a \frac{\partial T}{\partial y} + \langle v \rangle T \\ F_q &= -K_q \frac{\partial q(T)}{\partial y} + \langle v \rangle q(T) \end{aligned} \tag{3.17}$$

where K_o, K_a and K_q are all functions of latitude, ϕ, $q\ (T)$ is the water vapour mixing ratio, $\langle v \rangle$ is the zonally averaged wind speed and y is the distance in the poleward direction.

Obviously the more complicated the parameterization the more realistic one might claim the model to be. These are the two basically different methods of incorporating the div(F) term: the Newtonian form developed by Budyko (equation (3.12)) or the eddy diffusive mixing formula developed by Sellers (equation (3.16)). The choice is always to weigh the extra detail such as is offered by Sellers against the decreased computational time of Budyko's method.

3.4 A *BASIC* ENERGY BALANCE CLIMATE MODEL

This type of climate model is a useful teaching/learning tool. The program listed in Figure 3.2 was written for undergraduate use at the University of Liverpool. It takes approximately 10k of memory.

The formulation of the EBM has been kept as simple as possible. The equations are those described in Section 3.2. A similar EBM is described in Appendix B but there different applications are discussed. The albedo parameterization is a simple 'on-off' step function based on a specified temperature threshold (see equation (3.10)). The emitted longwave radiation is a linear

```
5  REM ''''''''''''''''''''''''''''''''''''''''''''''''''''''''''''''''''
10 REM Energy Balance Model 1986 K.McGuffie & A. Henderson-Sellers
12 REM The model is designed to run on an IBMPC and does.
14 REM Other machines may require some alterations. Here CLS=clear screen,
16 REM . is used for formatted output, LPRINT sends output to a
18 REM printer. The code is designed to be as transparent as possible.
19 REM Unclosed quotes in output statements may not work on some machines.
20 FOR I = 1 TO 9
30 READ LATZ$(I)
40 NEXT I
50 DATA "80-90","70-80","60-70","50-60","40-50","30-40","20-30","10-20"," 0-10"
60 E$=CHR$(27):CLS
70 C=3.8
80 Q=342.5
90 A=204
100 B=2.17
110 IN= 3.14159/18!
120 PIBY2=3.14159/2!
130 DEF FNR(X)=INT(100*X)/100
140 FOR LAT = 1 TO 9
150 READ S(LAT)
160 NEXT LAT
170 DATA 0.5,0.531,0.624,0.77,0.892
180 DATA 1.021,1.12,1.189,1.219
190 PRINT
200 PRINT
210 PRINT
220 PRINT "*********************************************************************
230 PRINT "                              ENERGY
240 PRINT "                              BALANCE
250 PRINT "                              MODEL
260 PRINT "*********************************************************************
270 PRINT
280 PRINT "          Copyright      A. Henderson-Sellers
290 PRINT "                           & K. McGuffie
300 PRINT "                                                  (1986)
310 PRINT
320 PRINT "    This model is similar to those of Budyko and Sellers.
330 PRINT "    You will be offered the opportunity to alter the
340 PRINT "    values of the parameters which control the model climate.
350 PRINT "
360 PRINT
370 PRINT "
380 PRINT "              Press the space bar to continue
390 GOSUB 3120
400 CLS
410 PRINT
420 PRINT "         There are various possibilities for changing
430 GOSUB 3160
440 PRINT "         the model climate.     You can then test the
450 GOSUB 3160
460 PRINT "         sensitivity of this climate to changes in the
470 GOSUB 3160
480 PRINT "         solar constant.    That you should observe the
490 GOSUB 3160
500 PRINT "         changes due to your changing of the model
510 GOSUB 3160
520 PRINT "         parameters is also of importance in understanding
530 GOSUB 3160
540 PRINT "         the nature of this model.
550 PRINT
560 PRINT
570 PRINT "         Press space to continue "
580 GOSUB 3120
590 FOR I = 1 TO 9
600 READ ALAND(I)
610 NEXT I
620 DATA 0.5,0.5,0.452,0.407,0.357,0.309,0.272,0.248,0.254
630 AICE=.62
640 TCRIT=-10
650 C=3.8
660 Q=342.5
```

Figure 3.2 BASIC computer code for the energy balance climate model. The code was designed for an IBM PC-type computer.

```
670 A=204
680 B=2.17
690 CLS
700 PRINT "                 * * * * * * * * * * * * * * * * * * * *
710 PRINT "                          M A I N   M E N U
720 PRINT "                 * * * * * * * * * * * * * * * * * * * *
730 PRINT
740 PRINT
750 PRINT "         There are 3 main parameterization schemes within the
760 PRINT "         model.  You may make alterations to any or all of them
770 PRINT "         at any one time.  Any which you choose not to alter
780 PRINT "         will be filled by default values.
790 PRINT
800 PRINT
810 PRINT "   Choice                       Parameterization
820 PRINT "-------------------------------------------------------------------------------
830 PRINT "    (1)                          ALBEDO
840 PRINT "    (2)                          LATITUDINAL TRANSPORT
850 PRINT "    (3)                          LONGWAVE RADIATION TO SPACE
860 PRINT
870 PRINT "    (4)                          RUN
880 PRINT
890 PRINT "  Enter the number of your choice
900 N$=INKEY$
910 IF N$="1" THEN GOTO 960
920 IF N$="2" THEN GOTO 1600
930 IF N$="3" THEN GOTO 1800
940 IF N$="4" THEN GOTO 2020
950 GOTO 900
960 CLS
970 PRINT "* * * * * * * * * * * * * * * * * * * * * * * * * * * * * * * * *
980 PRINT "              P A R A M E T E R I Z A T I O N   O F
990 PRINT "                      A L B E D O
1000 PRINT "* * * * * * * * * * * * * * * * * * * * * * * * * * * * * * * *
1010 PRINT
1020 PRINT
1030 PRINT "        There are three things which you may alter
1040 PRINT
1050 PRINT "          1.   The temperature at which the surface "
1060 PRINT "               becomes ice covered."
1070 PRINT
1080 PRINT "          2.   The albedo of this ice covered surface
1090 PRINT
1100 PRINT "          3.   The albedo of the underlying ground.
1110 PRINT
1120 PRINT "          4.   Return to main menu
1130 PRINT
1140 PRINT " Choose which one you want"
1150 AL$=INKEY$
1160 IF AL$="1" THEN GOTO 1210
1170 IF AL$="2" THEN GOTO 1280
1180 IF AL$="3" THEN GOTO 1350
1190 IF AL$="4" THEN GOTO 690
1200 GOTO 1150
1210 CLS
1220 PRINT
1230 PRINT
1240 PRINT "   The default value is -10 deg.C"
1250 PRINT
1260 INPUT "   What is the new value you want for TCRIT ? ",TCRIT
1270 GOTO 960
1280 CLS
1290 PRINT
1300 PRINT "   the albedo of the ice is currently 0.62."
1310 PRINT
1320 INPUT "   What is your new value for this albedo ? ",AICE
1330 IF AICE > .99 OR AICE < 0! THEN 1280
1340 GOTO 960
1350 CLS
1360 PRINT
1370 PRINT "        The albedos look like this from north to equator "
1380 PRINT
1390 PRINT "       (1)        80-90  ";ALAND(1)
```

Figure 3.2 (continued)

```
1400 PRINT "          (2)          70-80   ";ALAND(2)
1410 PRINT "          (3)          60-70   ";ALAND(3)
1420 PRINT "          (4)          50-60   ";ALAND(4)
1430 PRINT "          (5)          40-50   ";ALAND(5)
1440 PRINT "          (6)          30-40   ";ALAND(6)
1450 PRINT "          (7)          20-30   ";ALAND(7)
1460 PRINT "          (8)          10-20   ";ALAND(8)
1470 PRINT "          (9)           0-10   ";ALAND(9)
1480 PRINT
1490 PRINT "  Which one do you want to alter ( zero for none of them )"
1500 PRINT
1510 INPUT "   Enter the number ",I
1520 IF I = 0 THEN GOTO 960
1530 PRINT
1540 PRINT "   The old value in band ",I," is ",ALAND(I)"."
1550 INPUT "   What is your new value ? ",ALAND(I)
1560 IF ALAND(I) >0! AND ALAND(I) < 1! GOTO 1350
1570 GOSUB 3200
1580 GOTO 1550
1590 PRINT
1600 CLS
1610 PRINT
1620 PRINT "* * * * * * * * * * * * * * * * * * * * * * * * * * * * * * * * * * *
1630 PRINT "                  T R A N S P O R T
1640 PRINT "* * * * * * * * * * * * * * * * * * * * * * * * * * * * * * * * * * *
1650 PRINT
1660 PRINT "      In this case you can alter the rate at which heat is
1670 PRINT "      transported around the model by varying the value of C
1680 PRINT "      in the following equation.
1690 PRINT
1700 PRINT "             Heat Flux = C x ( T(mean) - T(zone) )"
1710 PRINT
1720 PRINT "                  The current value is ",C
1730 PRINT
1740 INPUT "   What is the value you want to use ? ", C
1750 IF C >0 AND C<50 GOTO 690
1760 GOSUB 3200
1770 GOTO 1740
1780 GOTO 690
1790 PRINT
1800 CLS
1810 PRINT "* * * * * * * * * * * * * * * * * * * * * * * * * * * * * * * * * * *
1820 PRINT "              L O N G W A V E  L O S S  T O  S P A C E
1830 PRINT "* * * * * * * * * * * * * * * * * * * * * * * * * * * * * * * * * * *
1840 PRINT
1850 PRINT "        The longwave loss to space is determined by the
1860 PRINT "        following equation.
1870 PRINT
1880 PRINT "            R = A + B x T(zone) "
1890 PRINT
1900 PRINT "      Currently        A=";A
1910 PRINT "                       B=";B
1920 PRINT
1930 PRINT "   Enter  1 to change them 0 to keep them the same"
1940 R$=INKEY$
1950 IF R$="0" THEN GOTO 690
1960 IF R$="1" THEN GOTO 1980
1970 GOTO 1940
1980 PRINT
1990 INPUT " Enter new value for A";A
2000 INPUT " Enter new value for B";B
2010 GOTO 690
2020 CLS
2030 PRINT
2040 PRINT
2050 PRINT "    What fraction of the solar constant would you like ?"
2060 INPUT "    Your choice >",SX
2070 REM start of routine to calculate temperatures
2080 FOR LAT = 1 TO 9
2090 READ TSTART(LAT)
2100 NEXT LAT
2110 DATA -16.9,-12.3,-5.1,2.2,8.8,16.2,22.9,26.1,26.4
2120 F=1
```

Figure 3.2 (continued)

```
2130 FOR LAT= 1 TO 9
2140 TEMP(LAT)=TSTART(LAT)
2150 NEXT LAT
2160 FOR H = 1 TO 50
2170 SOLCON=SX*1370!/4!
2180 REM Calculate albedo of zones
2190 LATICE=0
2200 FOR LAT = 1 TO 9
2210 NL=0
2220 ALBEDO(LAT)=ALAND(LAT)
2230 IF TEMP (LAT) > TCRIT THEN GOTO 2440
2240 ALBEDO(LAT) = AICE
2250 IF LAT = 9 THEN GOTO 2430
2260 IF TEMP(LAT+1)<=TCRIT THEN GOTO 2440
2270 DP=(-TCRIT+TEMP(LAT+1))*.1745/(TEMP(LAT+1)-TEMP(LAT))
2280 WORK2=(PIBY2-(LAT+.5)*IN)
2290 LATICE =WORK2+DP
2300 IF DP > .0872564 THEN GOTO 2370
2310 A3=PIBY2-(LAT+1)*IN
2320 A4=A3-IN
2330 A5=(SIN(A4)-SIN(LATICE))/(SIN(A4)-SIN(A3))
2340 NC=ALBEDO(LAT+1)*(1!-A5)+AICE*A5
2350 NL=LAT+1
2360 GOTO 2440
2370 A3=PIBY2-LAT*IN
2380 A4=PIBY2-(LAT-1)*IN
2390 A5= (SIN(LATICE)-SIN(A3))/(SIN(A4)-SIN(A3))
2400 NC=AICE-(AICE-ALBEDO(LAT))*A5
2410 NL=LAT
2420 GOTO 2440
2430 NL=0
2440 NEXT LAT
2450 IF ALBEDO(1) = ALAND(1) THEN NI = 90!/57.296
2460 REM CALCULATE MEAN TEMPERATURE
2470 SM=0
2480 FOR LAT = 1 TO 9
2490 WORK1=PIBY2-(LAT-1)*IN
2500 WORK2=WORK1-IN
2510 AC=ALBEDO(LAT)
2520 IF LAT=NL THEN AC=NC
2530 SM=SM+(SIN(WORK1)-SIN(WORK2))*AC*S(LAT)
2540 NEXT LAT
2550 TX=(SOLCON*(1-SM)-A)/B
2560 REM END OF MEAN TEMP ROUTINE
2570 FOR LAT = 1 TO 9
2580 TM(LAT)=TEMP(LAT)
2590 TEMP(LAT)=(SOLCON*S(LAT)*(1-ALBEDO(LAT))-A+C*TX)
2600 TEMP(LAT)=FNR(TEMP(LAT)/(C+B))
2610 NEXT LAT
2620 REM NOW TEST FOR CONVERGENCE
2630 AM=0
2640 IC=0
2650 FOR LAT= 1 TO 9
2660 MA=ABS(TEMP(LAT)-TM(LAT))
2670 IF MA > AM THEN AM =MA
2680 NEXT LAT
2690 IF AM < .01 THEN IC = 1
2700 IF IC = 1 THEN GOTO 2740
2710 NEXT H
2720 PRINT "  Model has failed to converge - - Think about your input  "
2730 END
2740 REM RESULTS
2750 CLS
2760 PRINT " ----------------------------------------------------------------
2770 PRINT "                                     R E S U L T S
2780 PRINT " ----------------------------------------------------------------
2790 PRINT "              Zone                  Temperature              Albedo
2800 PRINT "           =============           ===========              ======
2810 FOR LAT = 1 TO 9
2820 PRINT "             ";
2830 PRINT LATZ$(LAT);
2840 PRINT "          ";
2850 PRINT USING "£££.£" ;TEMP(LAT);
```

Figure 3.2 (continued)

```
2860 PRINT USING "                    £££.££";ALBEDO(LAT)
2870 NEXT LAT
2880 LATICE= FNR(LATICE*57.296)
2890 PRINT "        The ice edge is at " LATICE " degrees north."
2900 PRINT
2910 PRINT "        Input parameters "
2920 PRINT "   Fraction of solar constant";FNR(SX)
2930 PRINT "        A=";A;"      B=";B
2940 PRINT "        C=";C
2950 PRINT " Ice albedo= ";AICE;"  Changes at ";TCRIT;" Deg C"
2960 PRINT
2970 PRINT "   Press space bar to continue"
2980 GOSUB 3120
2990 CLS
3000 PRINT
3010 PRINT "             Do you want to try again ?"
3020 PRINT
3030 PRINT "          (1)      Reset all parameters
3040 PRINT "          (2)      Try again with same parameters
3050 PRINT
3060 RESTORE 600
3070 CH$=INKEY$
3080 IF CH$="1" THEN GOTO 590
3090 IF CH$="2" THEN GOTO 690
3100 GOTO 3070
3110 END
3120 SP$=INKEY$
3130 IF SP$=" " THEN GOTO 3150
3140 GOTO 3120
3150 RETURN
3160 FOR I=1 TO 700
3170 NEXT I
3180 PRINT
3190 RETURN
3200 PRINT "   Illegal response try again"
3210 RETURN
```

Figure 3.2 (continued)

function of the zonal surface temperature (see equation (3.11)) and the transport term is given by equation (3.12).

The following sections contain a brief summary of the model presented in Figure 3.2 and suggest some exercises which demonstrate the model's behaviour.

Description of the EBM

The model is governed by the equation originally devised by both Sellers and Budyko in 1969:

$$\text{(Shortwave in)} = \text{(Transport out)} + \text{(Longwave out)} \tag{3.18}$$

which is formulated as

$$S(\phi)\,\{1-\alpha(\phi)\} = K\{T(\phi)-\bar{T}\} + \{A + BT(\phi)\} \tag{3.19}$$

where K = the transport coefficient (here set equal to 3.80 W m^{-2} °C^{-1}),
$T(\phi)$ = the surface temperature at latitude ϕ,
$\bar{T}$ = the mean global surface temperature,
A and B are constants governing the longwave radiation loss (here taking values $A = 204.0$ W m^{-2} and $B = 2.17$ W m^{-2} °C^{-1}),
$S(\phi)$ = the mean annual radiation incident at latitude ϕ,
$\alpha(\phi)$ = the albedo at latitude ϕ.

Note that if the surface temperature at ϕ is less than $-10\ °C$ the albedo is set to 0.62. The solar constant in the model is taken as $1370\ W\ m^{-2}$.

Table 3.1 EBM simulation display showing input parameters and resultant equilibrium climate

A	B	K
$204\ W\ m^{-2}$	$2.17\ W\ m^{-2}\ °C^{-1}$	$3.81\ W\ m^{-2}\ °C^{-1}$
$TC = -10$	$AC = 0.62$	
Fract. of solar const. = 1		
Latitude	Temp.(°C)	Albedo
85	−13.5	0.62
75	−12.9	0.62
65	−4.8	0.45
55	1.8	0.40
45	8.5	0.36
35	16.0	0.31
25	22.3	0.27
15	26.9	0.25
5	27.7	0.25

The EBM is designed to be used to examine the sensitivity of the predicted equilibrium climate to changes in the solar constant. If the default values for the variables A, B, K and the albedo formulation are selected, an equilibrium climate which is quite close to the present-day situation is predicted for a fraction $=1$ of the solar constant. This equilibrium climate is given in Table 3.1.

Once this equilibrium value for an unchanged solar constant has been seen the user can modify the fraction of the solar constant prescribed and note the changes in the predicted climate. More importantly, the EBM permits the user to alter the albedo formulation, the latitudinal transport and the parameters in the infrared radiation term and examine the sensitivity of the modified model.

EBM model code

In the program listed in Figure 3.2, an equilibrium solution is achieved by iterating the calculation of each zonal T_i of equation (3.13). A maximum of 50 iterations is allowed. The snow-free albedo of the planet has been coded as latitude dependent.

The following exercises are useful examples of the types of experiments that can be undertaken.

Energy balance model exercises

Exercise 1. (a) Using the default values of albedo, K, A and B determine what *decrease* in the solar constant is required *just* to glaciate the Earth completely (ice edge at 0° N).

(b) Select some other values of A, B, K and the albedo formulation and repeat Exercise 1(a).

Exercise 2. (a) Various authors have suggested different values for the transport coefficient, K. For instance, Budyko (1969) originally used $K = 3.81$ and Warren and Schneider (1979) used $K = 3.74$. How sensitive is the model's climate to the particular value of K?

(b) Investigate the climate that results when using very small or very large values of K. How sensitive are these different climates to changes in the solar constant? Try and 'predict' how you think the model will behave before you perform the experiment.

Exercise 3. (a) Observations show that land will be totally snow-covered during winter for an annual mean surface temperature of 0 °C, and oceans totally ice-covered all year for a temperature of about −13 °C. The model specifies a change from land/sea to snow/ice at −10°C. Alter this 'critical' temperature and investigate the change in the climate and the climatic sensitivity to changing the solar constant.

(b) The albedo over snow-covered areas can vary within the limits of 0.5–0.8 depending on vegetation type, cloud cover and snow/ice condition. Investigate the sensitivity of the climate to changing the snow/ice albedo.

Exercise 4. (a) There have been many suggestions for the values of the constants A and B determining the longwave emission from the planet—some have been dependent on cloud amount. Budyko (1969) originally used $A = 202\ \mathrm{W\ m^{-2}}$ and $B = 1.45\ \mathrm{W\ m^{-2}\ {}^{\circ}C^{-1}}$. Cess (1976) suggested $A = 212\ \mathrm{W\ m^{-2}}$ and $B = 1.6\ \mathrm{W\ m^{-2}\ {}^{\circ}C^{-1}}$. How do these different constants influence the climate and its sensitivity?

(b) Holding A constant, just vary B and investigate the effect on the climate. What does a variation of B correspond to physically?

Exercise 5. Repeat Exercise 1 with the values of A, B, K and the albedo formulation which you believe are 'best' (i.e. most physically realistic for the present-day climate). Once the Earth is just fully glaciated begin to increase the fractional solar constant. Determine how much of an increase in the solar constant is required before the ice retreats from the equator. Do you understand the value?

Once the simple exercises described above have been completed and the model *understood*, it can be modified to run a sequence of timesteps or to produce plots of time-varying values. The next section describes some other types of experiments which can be conducted with EBMs similar to this.

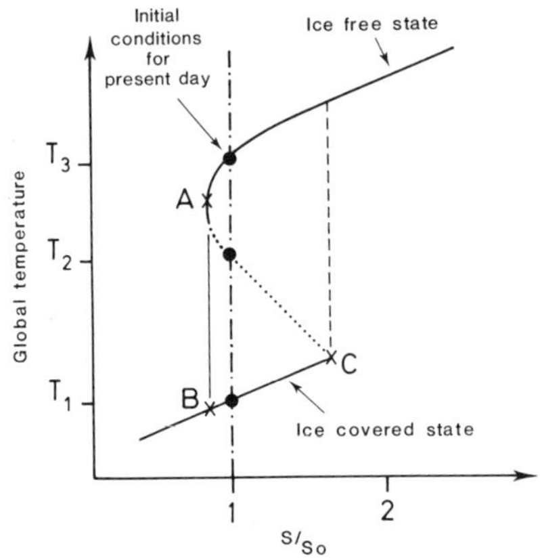

Figure 3.3 Characteristic solution of an EBM, plotted here as global mean temperature as a function of fraction of present-day solar constant. The dotted line represents a branch of the solution which, while being mathematically correct, is physically unrealistic. This lack of physical realism is due to the fact that on this branch increasing energy input results in a decreased temperature. More complex parameterizations within EBMs induce more complex shaped curves.

3.5 EXPERIMENTS WITH ENERGY BALANCE MODELS

So far we have looked at the components and the results of EBMs. In this section the results of some EBM experiments will be examined. In previous sections we have ignored seasonality and also have neglected the effect of the oceans as a heat source and sink. One common experiment is to introduce the effect of orbital (Milankovitch) variations on an EBM.

We will now consider in more detail the results of other more sophisticated models. We have already mentioned the rapid glaciation of the modelled Earth as a result of a decreased solar constant. The solution of the governing equation of such a model for various values of solar constant is shown in Figure 3.3. This is an illustration of the solution of a simple, zero-dimensional model. It shows the fundamental characteristic of non-linear systems. A slow decrease in the solar constant from initial conditions for the present day means a gradual decrease in temperature until the point is reached (point A) where a runaway feedback loop causes total glaciation and a rapid drop in temperature (solid line to point B). When the solar constant is then increased the process is not immediately reversed; the temperature follows a different route (dashed line) until at a value of the solar constant greater than that of the present day (point C) temperatures rise again. The modelled climate exhibits an hysteresis.

Energy balance models incorporate the cryosphere, which is the frozen water of the Earth, as if it were a thin, high-albedo covering of the Earth's surface. There are, however, some other important effects in the real climate system which are due to the dynamic behaviour of ice sheets. These are considered in the following section.

3.5.1 Explicit modelling of the cryosphere

No attempt is made in the formulation of energy balance climate models to differentiate between snow cover and glaciation in the albedo term. Ice sheets

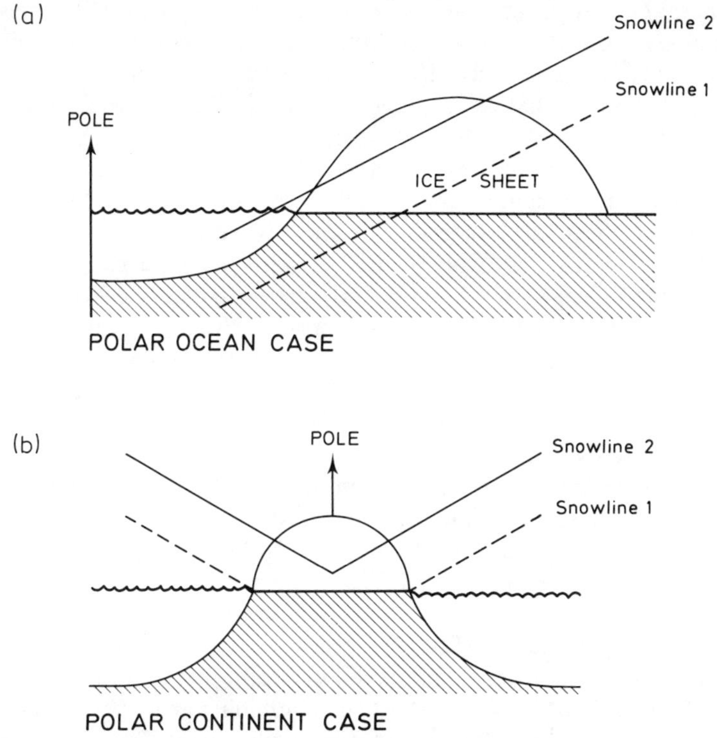

Figure 3.4 There are basically two different types of ice sheet: those occurring when there is a polar ocean and those which occur on a polar continent. In both cases it is possible for the ice sheet to persist even when the snowline is above ground level.

are, in reality, very different from transient snow cover and are complex. Figure 3.4 shows schematically two types of ice sheet. Figure 3.4(a) characterizes the major northern hemisphere ice sheets and Figure 3.4(b) characterizes the type of ice sheet which forms when a land mass exists at a pole, as is the case currently in the southern hemisphere.

Provided the snowline is below the level of the ice or bedrock topography then an ice sheet can exist. Once the ice sheet has acquired height then it can be sustained even if the snowline moves above the bottom of the ice sheet (i.e. snowline 1 changes to snowline 2). Such a situation can be found in Greenland at the present time. The extent of an ice sheet is governed by the balance between accretion and ablation. Accretion is by precipitation while ablation is the result of melting or calving (i.e. disintegration into the ocean).

An ice sheet model can be coupled to representations of the response of the lithosphere to ice load and to an EBM such as that described earlier in this chapter to make a combined model. Some models of this sort have been shown to exhibit internal variability. The components interact to form a constantly varying climate even without such external forcing as the Milankovitch variations.

Continental ice sheets and permafrost extent probably vary on timescales of approximately 1000–10 000 years, although shorter timescale effects have been suggested. The influence of ice sheets on global climate and in particular their influence on climatic change on geological timescales have been studied with the aid of simple models by several workers. The cryosphere incorporated in such EBMs can be described as a simple 'pancake' ice sheet which bears little resemblance to that which occurs naturally. It is important that the height and plasticity of ice sheets are taken into account when modelling them and their effects. These detailed cryospheric EBMs were being investigated by 1980. Sensitivity experiments showed that the influence of an ice sheet on the radiation balance was small if sea-ice and snow cover were already incorporated. On the other hand, the inclusion of the ice sheet height–accumulation feedback loop (the feedback which occurs when the height of the ice sheet increases and creates a greater area for accumulation by snowfall and, hence, a further increase in height) substantially increased climate sensitivity. This means that once the ice sheet has formed over a portion of the continent, additional accumulation forces the edge of the ice sheet further equatorward.

In modelling the response of ice sheets to Milankovitch variations a range of sensitivity experiments has shown that the precise outcome is highly dependent on the values of the input parameters. Using a coupled climate–ice sheet model, it can be shown that a satisfactory response to orbital variations requires consideration of bedrock deformation and incorporation of a calving mechanism to permit disintegration of the ice sheet. By combining an ice sheet model similar to that shown in Figure 3.4(a) with a two-dimensional EBM it is possible to simulate the glacial/interglacial cycles over the past 240 000 years. Although the ice sheet model simulates growth well it is found that the observed rapid dissipation of ice sheets can only be simulated by a parameterization of the calving and subsequent melting of icebergs. This in turn raises sea level and, by undercutting the ice sheets, forms further icebergs.

In the model of an ice sheet many different factors must be incorporated; the complexity of the formulation being related to the projected use of the model. If, for example, it was intended to try to simulate the Pleistocene ice sheets in Europe then only an estimation of the precipitation is required and it is also probably necessary to approximate the topography. However, when modelling the present-day ice-caps it is possible, and indeed essential, to utilize data gathered in field observational programmes.

3.6 BOX MODELS—ANOTHER FORM OF ENERGY BALANCE MODEL

The concept of computing the energy budget of an area or subsystem of the climate system can be extended and modified to produce other forms of EBMs. These models are not strictly EBMs and are often termed box models. A very elementary box model was considered early in this chapter (Section 3.2) in the example of the solar-heated swimming pool. That model had two boxes: one 'box' being the water and the other the air overlying the pool. A more complex consideration involves a more realistic parameterization of the energy transfer between the air and the pool, and interactive variation of other elements such as the radiative forcing. A simple column EBM has been used to consider the likely effect upon global temperatures of rising levels of atmospheric CO_2.

3.6.1 A simple box model of the ocean–atmosphere

The column EBM, used as an example here, represents the ocean–atmosphere system by only four 'compartments' or 'boxes': two atmospheric (one over land, one over ocean), an oceanic mixed layer and a deeper diffusive ocean. The heating rate of the mixed layer is calculated by assuming a constant depth in which the temperature difference ΔT, due to some perturbation, changes in response to: (i) the change in the surface thermal forcing, ΔQ; (ii) the atmospheric feedback, expressed in terms of a climate feedback parameter, λ, and (iii) the leakage of energy permitted into the underlying waters. This energy, ΔF, acts as a surface boundary condition to the deep ocean (i.e. the bottom of the mixed layer) in which the turbulent diffusion coefficient, K, is assumed to be a constant. The equations describing the rates of heating in the two 'layers'are thus:

(i) for the mixed layer

$$C_m \frac{d\Delta T}{dt} = \Delta Q - \lambda\, \Delta T - \Delta F \tag{3.20}$$

(ii) for the deeper waters

$$\frac{\partial\, \Delta T_0}{\partial t} = K \frac{\partial^2 \Delta T_0}{\partial z^2} \tag{3.21}$$

This latter equation may be evaluated at depth, z, (measured vertically downwards from zero at the interface) or calculated numerically using a vertical grid. In either case the heat source at the surface is the energy 'leaking' out of the mixed layer, ΔF, which thus acts as a surface boundary condition to the lower-level differential equation. However, a simpler parameterization can be utilized by assuming that at the interface there is continuity between mixed layer temperature change ΔT and deeper layer temperature change evaluated at the interfacial level, $\Delta T_0(0, t)$, i.e.

$$\Delta T_0(0, t) = \Delta T(t) \tag{3.22}$$

With this formulation, the value of ΔF can be calculated from

$$\Delta F = -\gamma \varrho c\, K \left\{ \frac{\partial\, \Delta T_0}{\partial z} \right\} \Bigg|_{z=0} \tag{3.23}$$

and used in equation (3.20). In this last equation, γ is the parameter utilized to average over land and ocean and has a value between 0.72 and 0.75, ϱ is the density of water and c its specific heat capacity.

The model described by equations (3.20) and (3.21) can be used to evaluate different atmospheric forcings, related to possible impacts of increasing atmospheric carbon dioxide. There are two possible forms for the change, ΔQ: either an instantaneous "jump"

$$\Delta Q = a \tag{3.24}$$

or a gradual increase

$$\Delta Q = bt\ \exp(\omega t) \tag{3.25}$$

Using both these possible forms for ΔQ, it is possible to compare a full numerical solution of the model with an approximation, which is gained by considering an infinitely deep ocean for which ΔF can be given by the expression

$$\Delta F = \gamma\mu\varrho ch \frac{\Delta T}{(\tau_d t)^{1/2}} \tag{3.26}$$

where μ is a tuning coefficient evaluated by comparison with the numerical solution, h is the mixed layer depth and $\tau_d (= \pi h^2/K)$, a characteristic time for exchange between the mixed layer and the deep ocean. Substituting equation (3.26) into equation (3.20) results in an ordinary differential equation:

$$\gamma \frac{\mathrm{d}\,\Delta T}{\mathrm{d}t} + \Delta T \left\{ \frac{1}{\tau_f} + \frac{\mu\gamma}{(\tau_d t)^{1/2}} \right\} = \frac{\Delta Q}{\varrho ch} \tag{3.27}$$

where $\tau_f = \varrho ch/\lambda$; which can then be solved analytically using a prescribed functional form for ΔQ. For the two expressions, given here as equations (3.24) and (3.25), values for the temperature increment over a period of 130 years (1850–1980) can be deduced (Figure 3.5) for chosen values of K and h. Here two sets of parameter values are shown. Using the CO_2 values observed for 1958 (315 ppmv) and 1980 (338 ppmv), the coefficients b and ω are easily evaluated from the equation for the increase of CO_2, which, corresponding to equation (3.25), is

$$C(t) = C_0 \exp(Bt\ \exp\ (\omega t)) \tag{3.28}$$

The two coefficients are determined by choice of initial (1850) CO_2 concentrations (horizontal axis in Figure 3.5), from which the coefficient b in equation (3.25) can be calculated as

$$b = \frac{B\ \Delta Q_{2x}}{\ln 2} \tag{3.29}$$

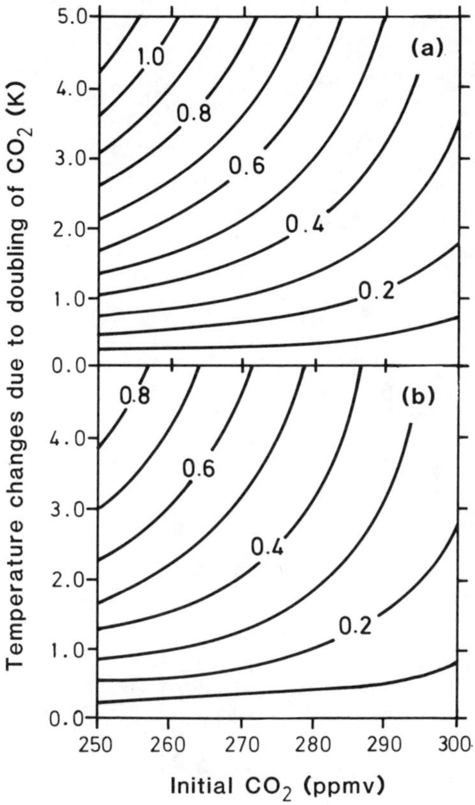

Figure 3.5 Isolines of temperature change to 1980 (CO_2 level of 338 ppmv) as a function of the CO_2-doubling temperature change and the 1850 initial CO_2 level for two pairs of ocean diffusivity and mixed layer depth: upper diagram, $K=10^{-4}\,m^2\,s^{-1}$, $h=70$ m; lower diagram, $K=3\times10^{-4}\,m^2\,s^{-1}$, $h=110$ m. Results are based on a full numerical solution of the equations described in Wigley and Schlesinger (1985). Reprinted by permission from *Nature*, **315**, 649–652. Copyright © 1985 Macmillan Journals Ltd.

where the atmospheric forcing resulting from a doubling of CO_2, ΔQ_{2x}, is related to the chosen values for the climate feedback parameter, λ, (where λ here is the same as λ_{TOTAL} defined in Section 1.4.4) and the assumed value for the CO_2 doubling temperature change, ΔT_{2x}, (vertical axis in Figure 3.5):

$$\Delta Q_{2x}=\lambda\ \Delta T_{2x} \tag{3.30}$$

From these diagrams it is apparent that for reasonable estimates of initial (viz. 1850 baseline) carbon dioxide concentration (270 ppmv), the expected 1850-to-date (1980) temperature increment of the mixed layer for a wide range (0–5 K) of expected temperature increments due to a doubling of CO_2 is well in accord with observations. (Note the observed air temperature increments must be assumed equal to the mixed layer temperature increases over the same period by assuming long-term quasi-equilibrium.) A numerical implementation of this simple box model is described in Appendix B.

3.6.2 A coupled atmosphere, land and ocean energy balance box model

It is possible to increase the level of complexity incorporated into a box model such as that described in the previous section, so that other features can be

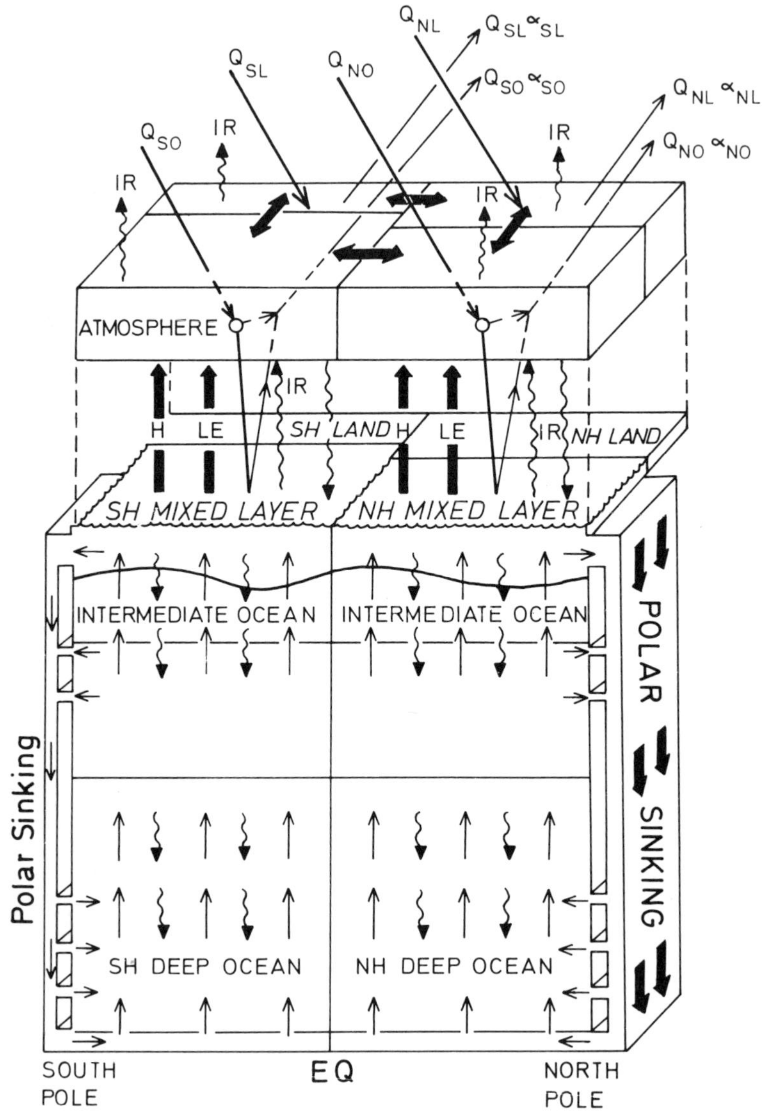

Figure 3.6 Illustration of the construction of and interactions within a complex box model of the Earth's climate system which includes hemispheric and land/ocean resolution and oceanic deep water formation (reproduced with permission from Harvey and Schneider, *J. Geophys. Res.*, **90**, 2207–2222 (1985), Copyright by the American Geophysical Union).

resolved. Figure 3.6 illustrates the components of an energy balance box model which includes separate subsystems for northern and southern hemisphere land, ocean mixed layer, ocean intermediate layer and deep oceans. This model separates the atmospheric response over land and ocean and incorporates polar

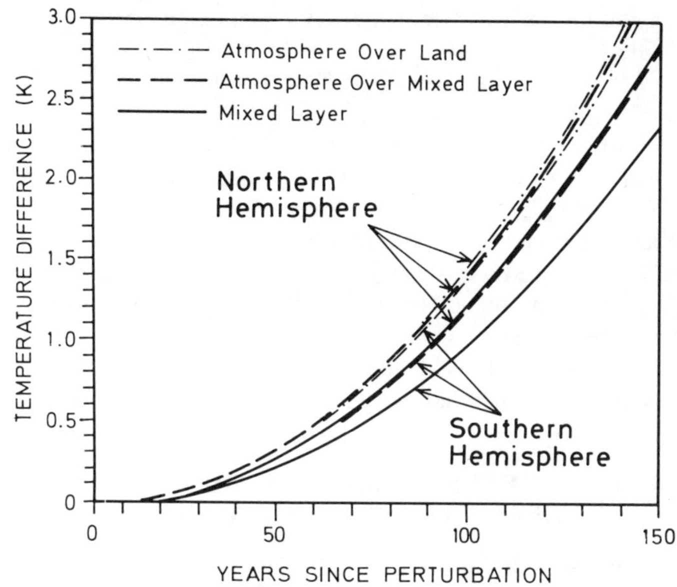

Figure 3.7 Effect of increasing CO_2 on the climate of the sophisticated box model of the climate system shown in Figure 3.6 (reproduced with permission from Harvey and Schneider, *J. Geophys. Res.*, **90**, 2207–2222 (1985), Copyright by the American Geophysical Union).

sinking of oceanic water into the deep ocean (the formation of deep bottom water). Despite these features, the model is essentially a box advection–diffusion model although it includes seasonally varying mixed layer depth and is forced with a seasonally varying insolation. As with all relatively simple models some features are prescribed. For example, hemispherically averaged cloud fraction is prescribed as a seasonally varying feature. As the land is hemispherically averaged, there is no opportunity to incorporate a temperature–surface albedo feedback in this sort of model.

Despite these constraints, this simple box model can be used to investigate sensitivity to features which have not yet been effectively incorporated into the coupled ocean–atmosphere GCMs such as those discussed in Chapter 6. For example, the response of atmospheric and mixed layer temperatures to feedback processes involving changes in vertical diffusivity and changes in vertical velocities can be computed explicitly. Figure 3.7 shows the response of the atmospheric temperatures over land and over the ocean, and of the oceanic mixed layer temperature of both hemispheres following a transient CO_2 perturbation simulated by a change in the parameterization of the infrared (IR) emission to space, where

$$\text{Emitted IR} = A' + B' \ T + (\text{cloud term}) \tag{3.31}$$

and the transient increase in atmospheric CO_2 causes a change in A' given by

$$\Delta A'(t) = -2.88 \times 10^{-4} t^2 \tag{3.32}$$

where t is the time, in years, since 1925.

In this model oceanic vertical velocities can change in perturbed climatic states. The results in Figure 3.7 follow from the velocity increase in the northern hemisphere and the decrease in the southern hemisphere. There is a faster mixed layer warming which reduces the lag of the mixed layer warming behind the atmospheric warming in the northern hemisphere as compared with the response in the southern hemisphere. These results suggest that more detailed analysis of oceanic feedback effects is required than can apparently be accomplished at present by three-dimensional coupled ocean–atmosphere models.

3.7 ENERGY BALANCE MODELS: DECEPTIVELY SIMPLE MODELS

In this chapter we have intentionally emphasized the simple basis of EBMs: the energy fluxes into and out of the planet as a whole (and of a region or climate subsystem) must balance unless there is cooling or heating. This concept is fundamental to climate modelling. It will recur in Chapter 4 in which the heating rates of each atmospheric layer are computed directly from the energy balance of the layers; in Chapter 5 in which the large scale flow between zones is given a physical basis; and in Chapter 6 where each modelled component in a GCM is seen to be driven by its energy budget.

The other topic which has been stressed in this chapter is computing. We wanted to underline that the basis of practically all climate modelling is (relatively) simple mathematical formulations and parameterizations represented in and executed by very fast computers. We have taken space to list the full code of one EBM in Figure 3.2. The code of a GCM written in a similar high-level language (most of them are currently written in FORTRAN which is fairly similar to BASIC) would be as thick as a substantial dictionary. More sophisticated GCMs have codes whose page listings are thicker than a stack of encyclopaedias but, despite this apparent complexity, the tutorial questions posed in this chapter and in Appendix B could usefully be considered with reference to more complex models. It is helpful to keep in mind the fundamental concepts developed in this chapter and to return, often, to the deceptively simple basis of the models described.

RECOMMENDED READING

Budyko, M. I. (1969) The effect of solar radiation variations on the climate of the Earth, *Tellus*, **21**, 611–619.

Cess, R. D. (1976) Climatic change, a reappraisal of atmospheric feedback mechanisms employing zonal climatology, *J. Atmos. Sci.*, **33**, 1831–1843.

Hansen, J., Russell, G., Lacis, A., Fung, I., Rind, D., and Stone, P. (1985) Climatic response times: dependence on climate sensitivity and ocean mixing, *Science*, **229**, 857–859.

Harvey, L. D. D., and Schneider, S. H. (1985) Transient climate response to external forcing on 10^0–10^4 year timescales. 2. Sensitivity experiments with a seasonal, hemispherically averaged, coupled atmosphere, land, and ocean energy balance model, *J. Geophys. Res.*, **90**, 2207–2222.

North, G. R., Cahalan, R. F., and Coakley, J. A. (1981) Energy balance climate models, *Rev. Geophys. Space Phys.*, **19**, 91–121.

Oerlemans, J., and van der Veen, C. J. (1984) *Ice Sheets and Climate*, Reidel, Dordrecht, 217pp.

Sellers, W. D. (1969) A global climatic model based on the energy balance of the Earth–atmosphere system, *J. Appl. Met.*, **8**, 392–400.

Warren, S. G., and Schneider, S. H. (1979) Seasonal simulation as a test for uncertainties in the parameterization of a Budyko–Sellers zonal climate model, *J. Atmos. Sci.*, **36**, 1377–1391.

Wigley, T. M. L., and Schlesinger, M. E. (1985) Analytical solution for the effect of increasing CO_2 on global mean temperature, *Nature*, **315**, 649–652.

Chapter 4

Radiative–Convective Models

'Who can number the clouds in wisdom?'
Job 38: 37

4.1 THE CONCEPT OF A RADIATIVE–CONVECTIVE CLIMATE MODEL

In earlier chapters the importance of the greenhouse effect was noted. This effect is due to the absorption of the upwelling thermal infrared radiation which has been emitted by the surface of the Earth. If none of the gases in the Earth's atmosphere possessed absorption features in the wavelength region in which the Earth emits radiation, there would be no greenhouse addition and the surface temperature would be equal to the planetary effective radiative temperature (see Figure 3.1).

The greenhouse absorbers not only affect the surface temperature; they also modify the atmospheric temperature by their absorption and emission of radiation. Radiative–convective climate models are one-dimensional models like the EBMs described in Chapter 3. In this case, however, the dimension is the vertical. These models resolve many layers in the atmosphere and seek to compute the atmospheric and surface temperatures.

It is relatively simple to illustrate how RC models work. It is possible to derive a temperature profile for the atmosphere by dividing it into a number of layers. Suppose we divide the Earth's atmosphere into two layers so that each layer just absorbs the infrared radiation incident on it and, for this very simple example, let each layer be described as having an "infrared optical thickness of $\tau = 1$." The principal absorber in the Earth's atmosphere is water vapour which is contained within the lowest few kilometres. Thus our two layers can be shown to be centred at heights of 3 km (layer 1) and 0.5 km (layer 2). The infrared energy fluxes are shown in Figure 4.1(a). Both layers radiate as black bodies upwards and downwards and the ground radiates up. Since the planet is emitting at its effective temperature, T_e, and because all radiation from below is absorbed by the top layer, σT_e^4 must equal σT_1^4 and thus $T_1 = T_e$. The energy balance of the first atmospheric layer is (emitted = absorbed) or

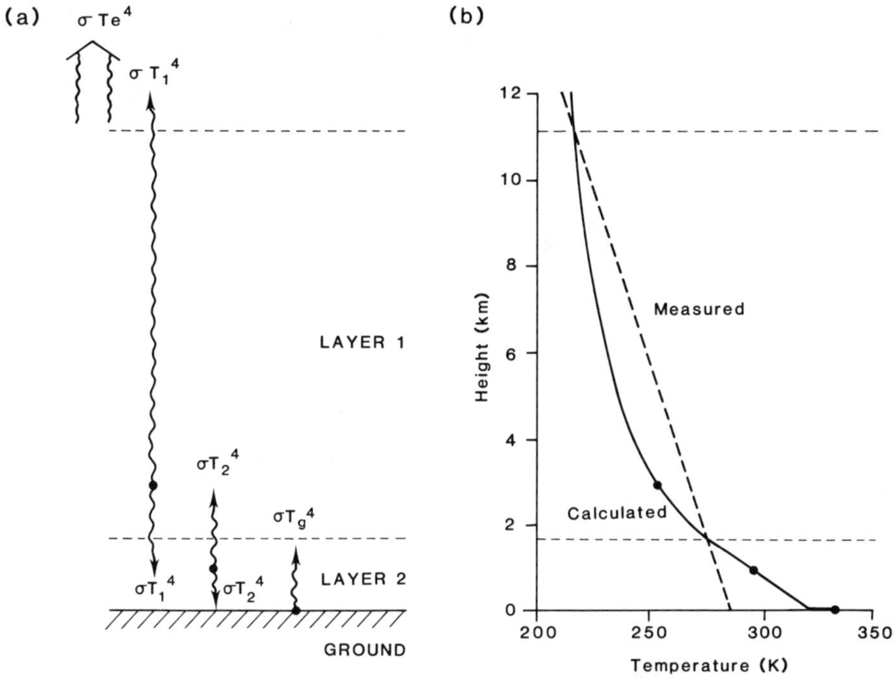

Figure 4.1 (a) The infrared radiative energy fluxes in the simple, two-layer model described. (b) The radiative equilibrium temperature profile calculated using the very simple model described in the text compared with a lapse rate of 6.5 K km^{-1}. This lapse rate is achieved by convection since the radiative equilibrium profile is unstable. (Richard M. Goody and James C. G. Walker, *Atmospheres*, ©1972, p. 56. Reprinted by permission of Prentice-Hall, Inc., Englewood Cliffs, New Jersey).

$\sigma T_2^4 = 2\sigma T_1^4 = 2\sigma T_e^4$. In general the temperature of layer n can be shown to be related to the effective temperature by

$$T^4(n) = \tau_{TOTAL}(n)\ T_e^4 \qquad (4.1)$$

where $\tau_{TOTAL}(n)$ is the total infrared optical thickness from the top of the atmosphere to layer n. In our simple case, because each layer has $\tau = 1$, $\tau_{TOTAL}(n) = n$. From equation (4.1) we can now calculate that the temperature of the top layer is equal to the effective temperature (255 K) and the lower-layer temperature is 303 K. The surface temperature can be obtained by considering the radiative budget of the second atmospheric layer. Here $2\sigma T_2^4 = \sigma T_g^4 + \sigma T_1^4$ so that $T_g = (3T_e^4)^{0.25}$, which is 335 K.

These approximations are drastic but the resultant temperature/height profile (Figure 4.1) is adequately close to the radiative equilibrium profiles which were first produced by RC models. Note that the surface air temperature is lower than the ground temperature and the observed (measured) lapse rate (change of temperature with height) is about 6.5 K km^{-1}. Compared with this standard

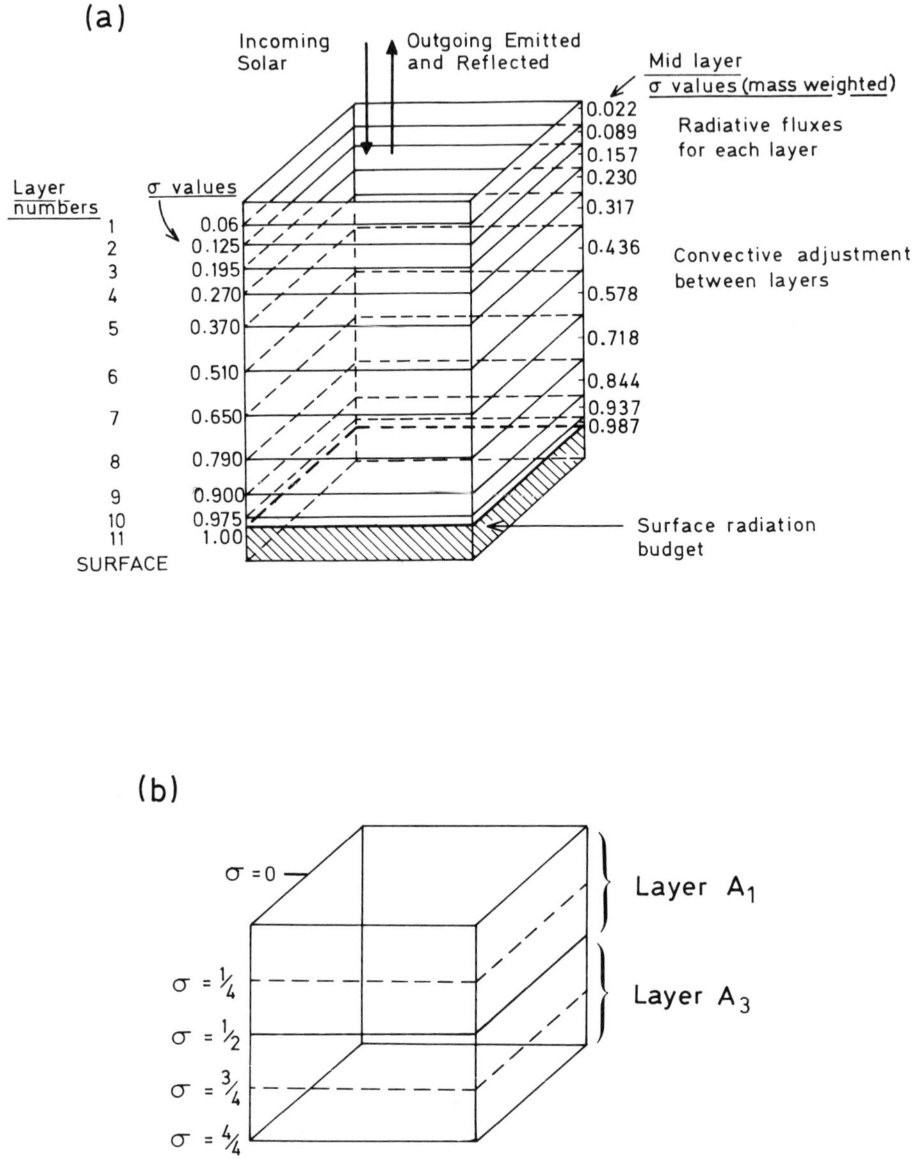

Figure 4.2 (a) The vertical structure of an RC model. This particular model has eleven layers arbitrarily placed to provide satisfactory vertical resolution near the surface and near the tropopause; (b) an illustration of the way in which σ (sigma) levels are constructed in a simpler case. $\sigma = 4/4$ is the surface. Shortwave and longwave radiative fluxes and hence the net radiative cooling are often calculated at the mid-points of the layers (i.e. at $\sigma = 1/4$ and $\sigma = 3/4$) and thus two radiative layers are represented. These can take their labels from their central σ levels i.e. layers A_1 and A_3. This labelling convention is used throughout Section 4.3.

(i.e. observed) lapse rate, our calculated radiative temperature profile is unstable. Thus if a small parcel of air were disturbed from a location close to the surface it would tend to rise because it would be warmer than the surrounding air. Its temperature would decrease at (roughly) the observed lapse rate so that at some arbitrary height (H) its temperature would be greater than that of the atmosphere. It would therefore continue to rise. Such a rising parcel of air would carry energy upwards and the resulting convection currents would mix the atmosphere, in this example throughout the whole depth of the troposphere. The convective mixing would alter the temperature profile until the atmosphere was dynamically stable.

This convective adjustment of a radiatively produced temperature profile is the essence of RC models. In the above highly simplified discussion we have made several gross assumptions but the basic concept of a radiatively computed temperature profile being adjusted to stability by parameterized convection is sound.

4.2 THE STRUCTURE OF GLOBAL RADIATIVE–CONVECTIVE MODELS

The RC model is a single column containing the atmosphere and bounded beneath by the surface. This bounded column usually represents the globally averaged conditions in the Earth–atmosphere system. As the name of the model type indicates, radiation and convection are treated explicitly. The radiation scheme is detailed and occupies the vast majority of the total computation time while the 'convection' is accomplished by a numerical adjustment of the temperature profile at the end of each timestep. Some RC models also include cloud prediction schemes but these are uncommon.

Figure 4.2 shows the major features of any RC model. The atmosphere is divided into a number of layers not necessarily of equal thickness. Eleven and four layers are illustrated but any reasonable number can be used. The layering can be defined with respect to height or pressure but it is more common to introduce the non-dimensional vertical coordinate, σ (sigma), which is a function of pressure

$$\sigma = \frac{p - p_{\mathrm{T}}}{p_{\mathrm{s}} - p_{\mathrm{T}}} \tag{4.2}$$

where p is the pressure, p_{T} the (constant) 'top-of-the-atmosphere' (usually a low stratosphere location) pressure and p_{s} the (variable) pressure at the Earth's surface. The top of the atmosphere has $\sigma = 0$ and the surface always has $\sigma = 1$. The σ values at each layer boundary in one such eleven-level atmosphere are given in Figure 4.2(a). In this example, which is taken from the UK Meteorological Office's 11-layer AGCM the top-of-the-atmosphere pressure is assumed to be 20 hPa and the surface pressure is roughly 1000 hPa. Intermediate levels which are the mid-points, in pressure coordinates, of the layers are also marked in Figure 4.2(a). It is at these levels that all the variables are calculated.

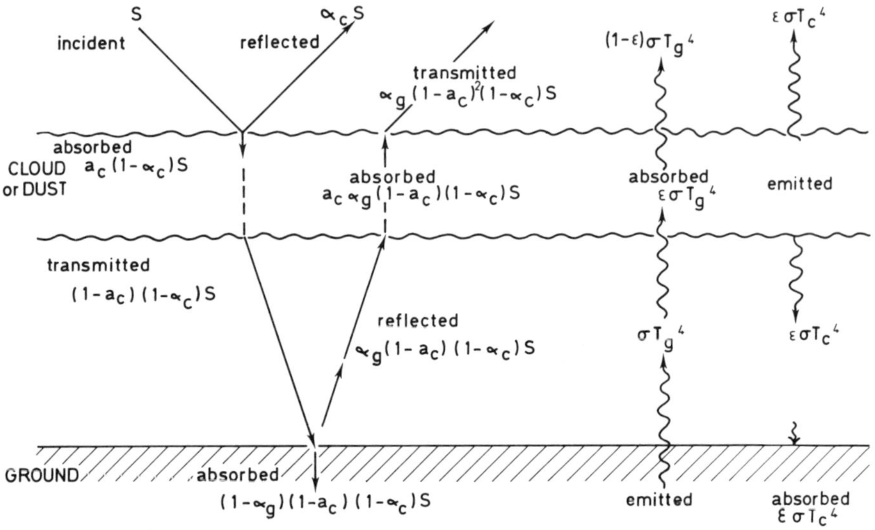

Figure 4.3 Idealized radiative interaction between a cloud or dust layer and the surface. There is assumed to be no reflection from the cloud base. The solar radiation suffers absorption on each passage through the layer. The infrared absorption is equal to the emissivity.

4.3 RADIATION COMPUTATION

The principles involved in radiative computations in climate models are most readily illustrated by considering a very simple global model in which a single cloud or aerosol layer is spread homogeneously over the surface (Figure 4.3). The incident solar radiation of S W m^{-2} can be traced as it interacts with the cloud. A part is reflected, a part absorbed and a part transmitted; the cloud albedo, α_c, and shortwave absorptivity, a_c, controlling these interactions. The remaining solar flux interacts with the surface. Here only reflection and absorption occur and the reflected ray can be followed through the cloud and out to space. The infrared radiation emitted from the surface is partially absorbed by the cloud. Here it is assumed that the infrared absorption of the cloud is equal to the infrared emissivity of the cloud, ϵ. The downwelling radiation from the cloud adds to the surface heating thus contributing to the greenhouse effect. Three main assumptions have been made in this simple model: (i) there is no reflection of the upward-travelling shortwave radiation by the cloud; (ii) the surface emissivity has been set equal to unity and (iii) the cloud/dust absorption in the infrared wavelength region is equal to ϵ. With all these assumptions, three energy balance equations can be written if it is also assumed that the model has reached equilibrium. Absorbed and emitted or reflected radiation at each level are equated so that

$$S = \alpha_c S + \alpha_g(1 - a_c)^2(1 - \alpha_c)S + \epsilon\sigma T_c^4 + (1 - \epsilon)\sigma T_g^4 \tag{4.3}$$

$$a_c(1-\alpha_c)S + a_c\alpha_g(1-a_c)(1-\alpha_c)S + \epsilon\sigma T_g{}^4 = 2\epsilon\sigma T_c{}^4 \tag{4.4}$$

$$(1-\alpha_g)(1-a_c)(1-\alpha_c)S + \epsilon\sigma T_c{}^4 = \sigma T_g{}^4 \tag{4.5}$$

Equations (4.3)–(4.5) represent the energy balances at the top of the atmosphere, the cloud level and the surface respectively. These equations can be solved directly by giving values for the dust/cloud shortwave absorption, a_c, albedo, α_c, its infrared emissivity, ϵ, and the surface albedo, α_g. Alternatively, the surface albedo term and the cloud temperature term can be eliminated from the equations leaving an expression for T_g

$$\sigma T_g{}^4 = \frac{S(1-\alpha_c)}{(2-\epsilon)}(2-a_c) \tag{4.6}$$

Taking the value of S as a quarter of the solar flux at the planet, i.e. $S = 343\ \text{W m}^{-2}$, and considering first the cloudless atmosphere case, if $\alpha_c = 0.08$, appropriate to scattering by atmospheric molecules alone, $a_c = 0.15$, representing absorption by the atmosphere, and $\epsilon = 0.4$, then the surface temperature is 283 K. This is quite close to the global average surface temperature of ~288 K.

We can now consider the addition of 'clouds' to this simple model; first a volcanic aerosol cloud and second a water droplet cloud. Inserting a volcanic aerosol into an otherwise cloudless atmosphere will increase the albedo slightly, say $\alpha_c = 0.12$, and will increase the solar absorption and the infrared emissivity, say to $a_c = 0.18$ and $\epsilon = 0.43$. Thus this volcanic aerosol 'cloud' gives rise to a cooling with the global mean temperature dropping to 280 K. (Note that Figure 1.13 shows a similar cooling caused by the introduction of a volcanic aerosol cloud into a more complex RC model.) Alternatively, we could consider the introduction of a water droplet cloud into the original cloudless atmosphere. If we assume that the cloud cover is to be approximately the same as that of the present day then α_c increases, say to 0.30, a_c increases slightly to 0.20 and ϵ increases to 0.90. With these values for partly cloudy skies the globally averaged surface temperature is 288 K. The introduction of this water droplet cloud has increased the calculated surface temperature because, with the radiative characteristics we have chosen, the greenhouse effect of the cloud is greater than the albedo effect. If, however, we alter the selected value of α_c very slightly so that $\alpha_c = 0.40$ and allow the other values to remain the same then the surface temperature becomes 277 K. The cloud albedo effect has 'beaten' the greenhouse effect this time. The sensitivity of this simple model's climate (as represented by the computed surface temperature) illustrates the interconnected role of various parameters in radiative transfer calculations and, in particular, the importance of the relative impact on the absorption of incoming solar radiation and the emission of infrared radiation.

The following treatment of atmospheric radiation is more complex than this stylized model but still somewhat simplified. The principles are the same in most models, the major complicating features being detailed consideration of the dependence of scattering and absorption on wavelength and other atmospheric variables.

The temperature profile of the atmospheric column is computed by calculating the net radiative heating in each layer. Calculations are made in terms of the layer potential temperature Θ_i given by

$$\Theta_i = T_i\ (p_0/p)^{\varkappa} \tag{4.7}$$

where $p_0 = 1000$ hPa is a reference pressure and $\varkappa = R/c_p = 0.286$, R is the gas constant for air and c_p is its specific heat at constant pressure.

To simplify the discussion of radiative transfer let us assume that there are only two layers, A_1 and A_3, in the atmosphere as shown in Figure 4.2(b). The radiation is treated as two separate streams. The shortwave radiation includes all the solar radiation, regardless of wavelength, and the attenuation of this radiation by Rayleigh scattering, its reflection from the Earth's surface and from clouds, and its absorption in the atmosphere and in clouds. The second stream is the flux of longwave radiation including all that which is emitted by the atmosphere, clouds and the Earth's surface. The ground temperature, T_g, needed in order to evaluate the evaporation, the sensible heat flux from the surface and the net longwave surface radiation, is determined from the heat balance at the Earth's surface.

4.3.1 Shortwave radiation

In this simple case we divide the incoming solar radiation into two parts by wavelength, the division being somewhere between 0.7 and 0.9 μm. The two wavelength regions can then either be treated separately but identically or the absorption and scattering can be partitioned by wavelength so that the two parts of the radiation are designated R_s (the shortwave part which is roughly 65 per cent of the total and is subject to Rayleigh scattering) and R_a (the near infrared wavelength part which is roughly 35 per cent of the total and is subject to atmospheric absorption). These can be approximated as

$$R_s = 0.65S\ \cos\mu \tag{4.8}$$

$$R_a = 0.35S\ \cos\mu \tag{4.9}$$

where S is the solar constant (adjusted for the Earth–Sun distance) and μ is the zenith angle of the Sun. Some models calculate a mean value of μ, which is used during the hours of daylight. A summary of the disposition of these components of the shortwave radiation for both clear and cloudy skies is given in Figure 4.4 and is described in detail in the following paragraphs.

Albedo

The albedo of the clear atmosphere for the portion of the radiation which is assumed to be subject to Rayleigh scattering is given by

$$\alpha_0 = \min[1,\ 0.085 - 0.247\ \log_{10}\{(p_0/p_s)\cos\mu\}] \quad (4.10)$$

For an overcast atmosphere, the albedo for the scattered part of the radiation is composed of the contributions of Rayleigh scattering (by atmospheric molecules) and of Mie scattering (by cloud droplets). The simplest useful formulation is

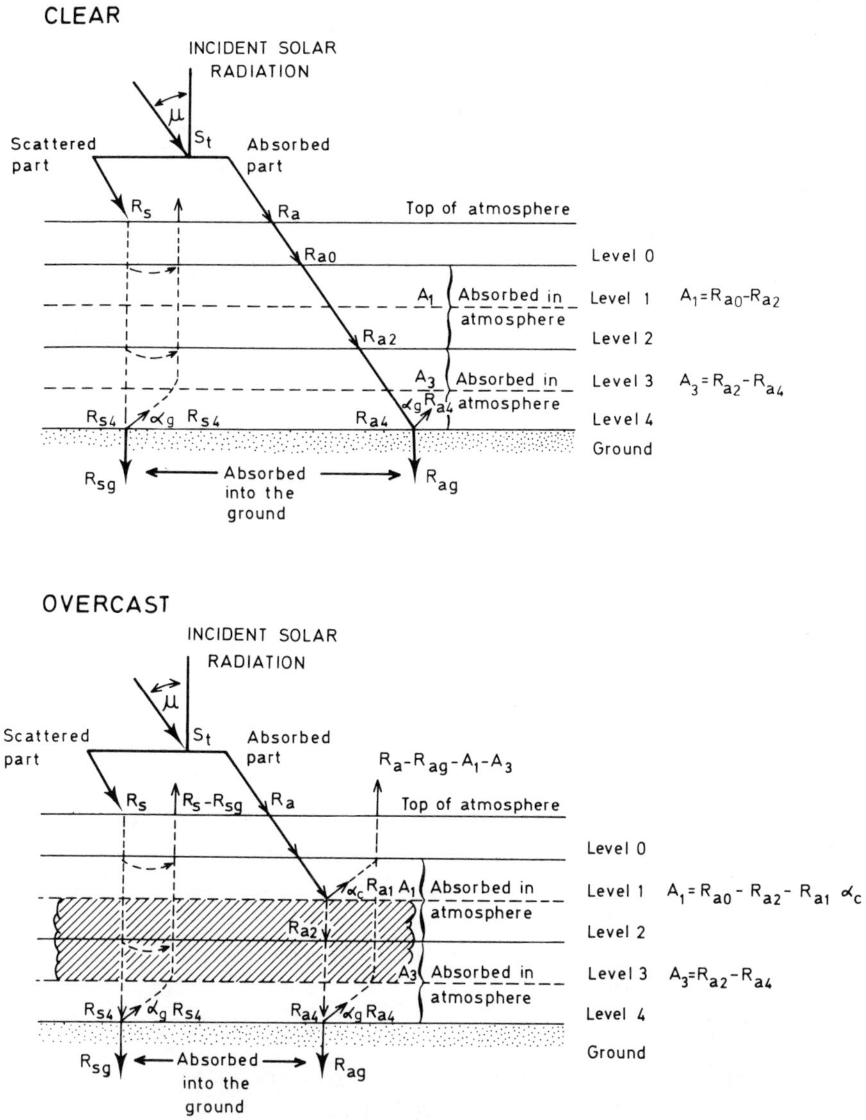

Figure 4.4 Schematic illustration of the principles of radiative transfer calculations. In this example there are only two atmospheric layers. More complex techniques involve many layers and many spectral intervals. All schemes consider only clear and cloudy cases. Intermediate cloud amounts are simulated by weighting these results.

$$\alpha_{ac} = 1 - (1 - \alpha_0)(1 - \alpha_c) \tag{4.11}$$

where α_c is the cloud albedo for both R_a and R_s (see Table 4.1).

The albedo of the surface is dependent upon surface types. A reasonable global average albedo for the entire solar spectrum is $\alpha_g = 0.10$. For specific surfaces it is often considered advantageous to introduce a spectral dependence since vegetation albedos increase rather sharply at around 0.7 μm while snow and ice albedos begin to decrease at about the same wavelength (Figure 4.5).

Table 4.1 Typical albedos and absorptivities of clouds

Cloud level	Albedo	Transmissivity	Absorptivity
High	~0.25	~0.75	0.005
Middle	~0.60	~0.38	0.02
Low	~0.70	~0.36	0.035

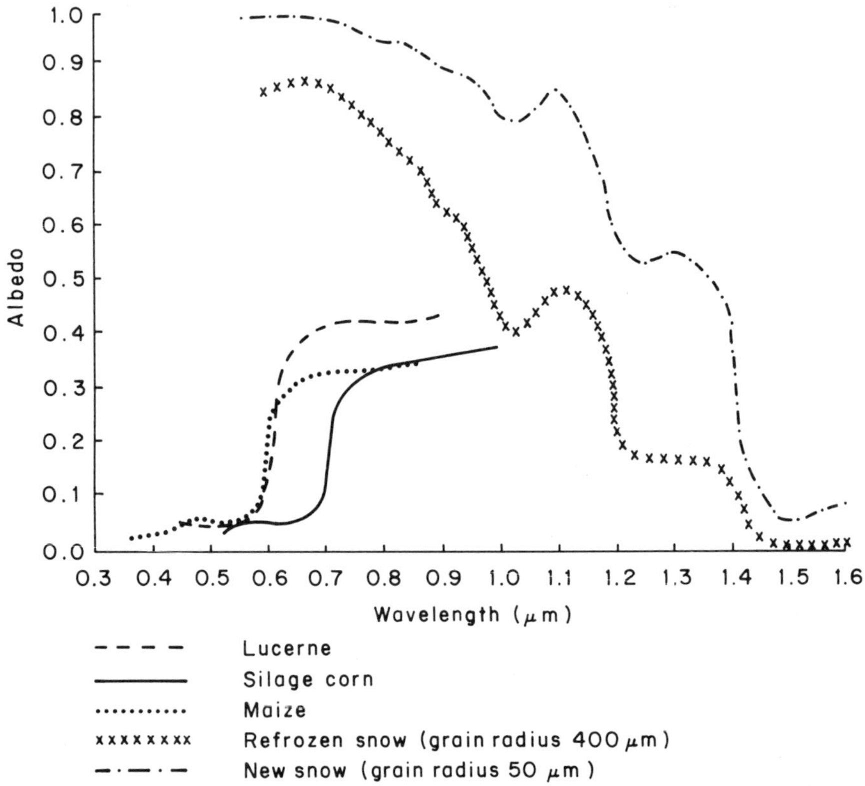

Figure 4.5 The spectral reflectance of different surface types. Note the markedly different spectral characteristics of snow surfaces and vegetated surfaces (reproduced with permission from Henderson-Sellers and Wilson, *Rev. Geophys. Space Phys.*, **21**, 1743–1778 (1983). Copyright by the American Geophysical Union).

Shortwave radiation subject to scattering (R_s)

The part of the solar radiation which is assumed to be scattered does not interact with the atmosphere, except to be partly scattered back to space. Thus the only part with which we are concerned is that amount which reaches, and is absorbed by, the Earth's surface. This is given by the expressions

$$R'_{sg} = R_s \frac{(1-\alpha_g)(1-\alpha_0)}{(1-\alpha_0\alpha_g)} \quad \text{for clear sky}$$

$$R''_{sg} = R_s \frac{(1-\alpha_g)(1-\alpha_{ac})}{(1-\alpha_{ac}\alpha_g)} \quad \text{for overcast sky} \tag{4.12}$$

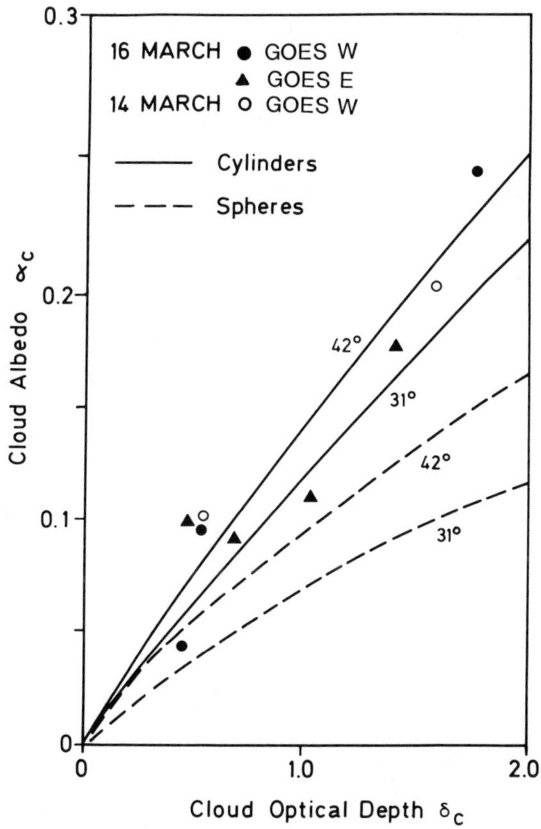

Figure 4.6 Cloud albedos, α_c, derived from the GOES satellite plotted as a function of lidar-derived cloud optical depths, δ_c. Also shown are theoretically predicted albedos, for two solar zenith angles, computed using ice spheres and ice cylinders (the latter being much more like the real cloud ice crystals) as the scattering objects (after Platt *et al.*, 1980).

Multiple reflections between sky and ground or between cloud base and ground are accounted for by the terms in the denominators. For partly cloudy conditions (neither clear nor overcast) the scattered radiation absorbed at the Earth's surface is

$$R_{sg} = NR''_{sg} + (1 - N)R'_{sg} \qquad (4.13)$$

where N is the fractional cloudiness of the sky. Cloud albedo is a function of both the character and thickness of the cloud as is illustrated by the experimental results shown in Figure 4.6. The absorption of this radiation by the ground affects the ground temperature and subsequently affects the longwave emission from the ground and the ground-level heat balance.

Shortwave radiation subject to absorption (R_a)

The solar radiation subject to absorption is distributed as heat to the various layers in the atmosphere and to the Earth's surface. The absorption depends upon the effective water vapour content as well as the ozone and carbon dioxide amounts. Usually these absorptivities are computed semi-empirically using formulae appropriate to wide spectral intervals.

For a cloudy sky the absorption in a cloud is generally prescribed as a function of cloud type only (see Table 4.1).

The incoming beam becomes diffuse within any cloud, and its path is assumed to be 1.66 times the vertical thickness of the cloud. The factor 1.66 is often termed the Elsasser factor and is derived by assuming that the diffuse radiation is isotropic. The mean path length is thus 1.66 times the vertical thickness of the layer. Below the cloud the beam is still diffuse, and the factor 1.66 for path length is retained.

When the sky is partly cloudy, the total flux at level i is given by a weighted average of the clear and overcast fluxes:

$$R_{ai} = NR''_{ai} + (1 - N)R'_{ai} \qquad (4.14)$$

That part of the flux subject to absorption which is actually absorbed by the ground is given (from Figure 4.4) by

$$R'_{ag} = (1 - \alpha_g)R'_{a4} \qquad (4.15)$$

for clear sky, and by

$$R''_{ag} = \frac{(1 - \alpha_g)R''_{a4}}{1 - \alpha_c\alpha_g} \qquad (4.16)$$

for completely cloudy (overcast) sky, where the factor $1/(1 - \alpha_c\alpha_g)$ again accounts for multiple reflections between the ground and cloud base. For partly cloudy skies, the radiation absorbed by the ground is therefore given by

$$R_{ag} = NR''_{ag} + (1 - N)R'_{ag} \tag{4.17}$$

The total solar radiation absorbed by the ground will be the sum of that part of the solar radiation spectrum which is subject to (atmospheric) absorption, that is absorbed by the ground, and that part of the spectrum subject to atmospheric scattering that is absorbed by the ground; thus giving

$$R_g = R_{ag} + R_{sg} \tag{4.18}$$

4.3.2 Longwave radiation

The calculation of the longwave radiation, like that of the shortwave radiation, is based on an empirical transmission function depending primarily upon the amount of water vapour. The net longwave radiation at any level can be expressed as the sum of two terms (Figure 4.7):

$$F_{(net)} = F\uparrow - F\downarrow \tag{4.19}$$

The upward flux at $z = h$ for radiation at some wavelength, λ, consists of the sum of two terms,

$$F_\lambda\uparrow(h) = B_\lambda[T(0)]\tau_\lambda(h,0) + \int_0^h B_\lambda[T(z)](\mathrm{d}/\mathrm{d}z)\tau_\lambda(h,z)\mathrm{d}z \tag{4.20}$$

The first term in equation (4.20) is the infrared flux arriving at $z = h$ from the surface. It is given by the surface flux $B_\lambda[T(0)]$ multiplied by the infrared transmittance of the atmosphere, τ_λ. The second term in equation (4.20) is the contribution to the total upward flux from emission of infrared radiation by atmospheric gases below the level $z = h$. Unlike the surface emission, which is nearly ideal black body radiation, the atmospheric emission as incorporated into the z derivative of τ_λ is highly wavelength-dependent. This is a result of the selective absorption by CO_2 or H_2O in certain spectral regions.

The downward infrared flux is composed solely of atmospheric emission, since the incoming planetary infrared radiation from space is essentially zero. The downwelling radiation at a layer at $z = h$ is given by

$$F_\lambda\downarrow(h) = \int_h^\infty B_\lambda[T(z)](\mathrm{d}/\mathrm{d}z)\tau_\lambda(h,z)\mathrm{d}z \tag{4.21}$$

The transmittance, τ_λ, of the atmosphere for infrared radiation in the wavelength interval (λ, $\lambda + \Delta\lambda$) and between atmospheric altitudes $z = h_1$ and $z = h_2$, can be expressed in terms of the optical thickness of the atmosphere in that wavelength region. The degree of detail used in the wavelength integration depends upon the computational power available and also on the type of

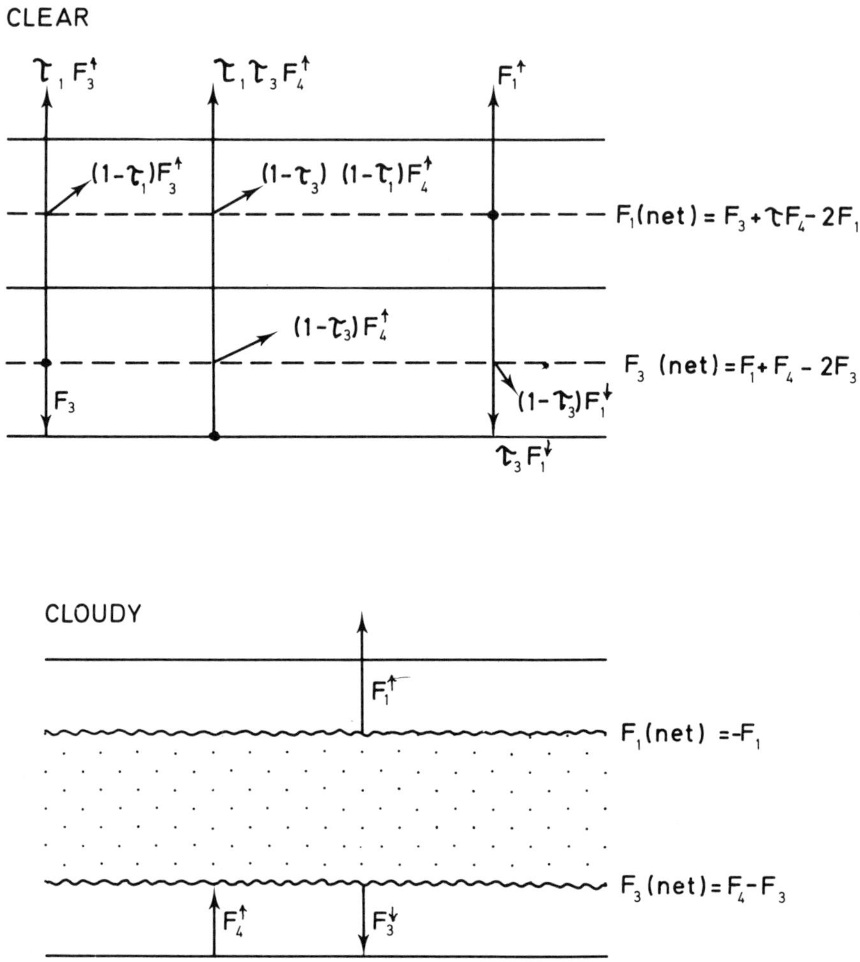

Figure 4.7 Schematic illustration of the transfer of terrestrial (longwave) radiation. There are fewer processes involved than for the case of shortwave radiation, but greater uncertainties in physical mechanisms make this portion just as difficult to deal with. (NB. Both the shortwave transfer (Figure 4.4) and the longwave interactions shown here are considerably simplified by the assumption of horizontally infinite clouds. No climate models yet incorporate finite clouds and cloud edge effects).

application. The availability of fast computer routines for performing certain mathematical calculations often makes it desirable to rewrite a problem such as this in an alternative form. There are two alternative methods of computing τ_λ using either a representation formulated in terms of exponential integrals or one involving proportionality to exponentials. Although the latter form is simpler to interpret there are fast numerical routines available to calculate the exponential integrals as rapidly as exponentials, and the former are more accurate. Such techniques are common in climate modelling and the development of fast, efficient algorithms is an important contribution to this subject.

For some sensitivity experiments where overlap between absorption spectra of different atmospheric constituents can affect the sensitivity of the model, detailed integration through at least some wavelength regions will be required. In other types of experiments, such as those concerned with cloud–radiation interactions for example, where the radiative fluxes are not strongly affected by the perturbed variable, band-averaging can be satisfactory. Band-averaging uses a wavelength (or more usually, frequency) averaged value of the Planck function, $B_\lambda(T)$, in equation (4.20) together with an averaged value of transmissivity. In this case the layer-averaged transmissivity τ_i can be calculated from the layer temperature and pressure and the amount of gaseous absorber.

Generally other uncertainties in the model, especially those generated by the somewhat arbitrary nature of the convective adjustment, will be greater than those associated with the difference between band-averaging or wavelength integration in the infrared calculations.

4.3.3 Heat balance at the ground

The ground temperature, T_g, is obtained from the heat balance at the ground. The treatment of the heating of the ground usually depends upon the assumed character of the ground or underlying surface. The albedo is wavelength-dependent and can also be dependent upon the surface moisture and vegetation. The ground can either be considered to be a perfect insulator with zero heat capacity or a heat capacity can be specified. The total flux of heat across the air/ground interface is given by

$$R_g + F - \epsilon\sigma T_g^4 - H_L - H_s = \text{stored energy} \tag{4.22}$$

where H_s is the sensible heat flux from the surface, H_L is the flux of latent heat due to evaporation from the surface, R_g is the solar radiation absorbed by the ground and F is the downwelling longwave radiation at the surface.

4.4 CONVECTIVE ADJUSTMENT

The computational scheme described so far defines a radiative temperature profile, $T(z)$, determined solely from the vertical divergence of the net radiative fluxes. Computation of globally averaged vertical radiative temperature profiles for clear sky conditions and with either a fixed distribution of relative humidity or a fixed distribution of absolute humidity yields very high surface temperatures, and a temperature profile which decreases extremely rapidly with altitude (i.e. despite the assumption made about the humidity distribution the computed radiative temperature profiles $T(z)$ have large lapse rates in the lower troposphere, considerably in excess of the mean value given by the US standard [McClatchey] atmosphere which has a lapse rate $\gamma_c = 6.5\,\text{K km}^{-1}$, e.g. Figure 4.1). This

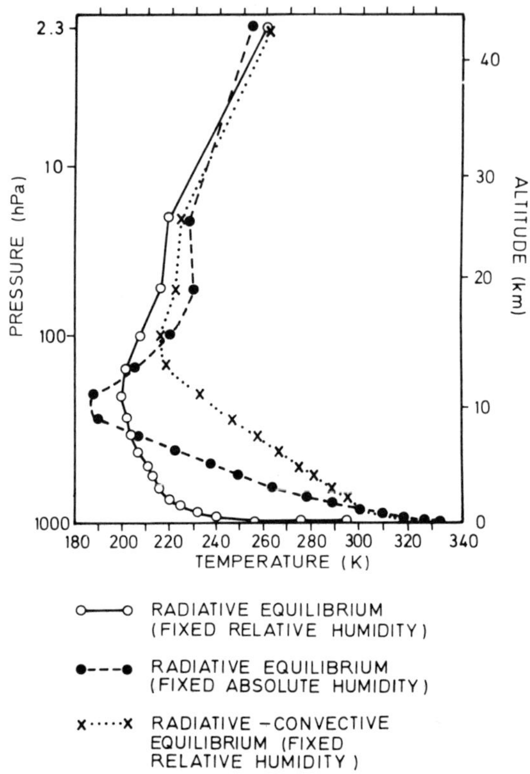

Figure 4.8 Convective adjustment. The solid line shows a calculated radiative equilibrium profile for a clear atmosphere with fixed relative humidity; the dashed line shows a calculated radiative equilibrium profile for a clear atmosphere with fixed absolute humidity; and the dotted line shows the RC equilibrium for the clear atmosphere with fixed relative humidity (reproduced by permission of the American Meteorological Society from Manabe and Wetherald, (1967), *Journal of the Atmospheric Sciences*, **24**, 241–259).

observed (critical) lapse rate γ_c is less than the computed $-\partial T/\partial z$ in the lower troposphere because the radiative equilibrium profiles are modified by free and forced vertical (moist and dry) convection and the vertical heat transport due to large scale eddies. Radiative equilibrium profiles are unstable to vertical (moist and dry) convection. Thus the tendency of radiative heating alone to produce large lapse rates adjacent to the Earth's surface, as shown by the solid and dashed curves in Figure 4.8, is offset by a rapid vertical transfer of heat.

By the mid-1960s it was realized that if column models were to produce meaningful values of surface and vertical temperatures it was necessary that the computed unstable profiles be modified. This modification is termed the 'convective adjustment'. It must be noted that it is *not* a computation of convection but rather a numerical re-evaluation which is applied whenever the critical lapse rate γ_c is exceeded in the time evolution of the numerical

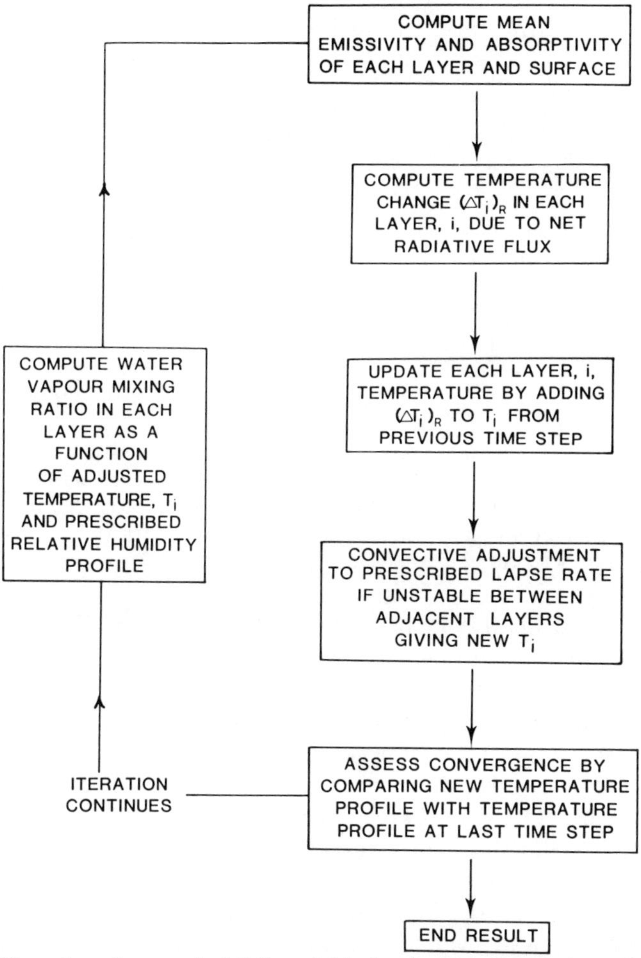

Figure 4.9 Flow chart for a typical RC model (subscript R indicates the radiative result). The model *must* satisfy the following conditions: (1) at the top of the atmosphere, net incoming solar radiation *must* equal net outgoing infrared radiation; (2) a specified vertical distribution of relative humidity is maintained; (3) no temperature discontinuity may exist between any two layers; (4) the lapse rate between two layers may not exceed the specified critical lapse rate (often 6.5 K km^{-1}); (5) if the lapse rate is less than this, local radiative equilibrium is satisfied.

calculation. The temperature difference between vertical layers is adjusted to the critical lapse rate γ_c by changing the temperature with time according to the integrated rate of heat addition. It can be shown that, if continuity of temperature across the radiation/convection interface is satisfied at one time in the course of a time-dependent calculation, it will be satisfied for all later times.

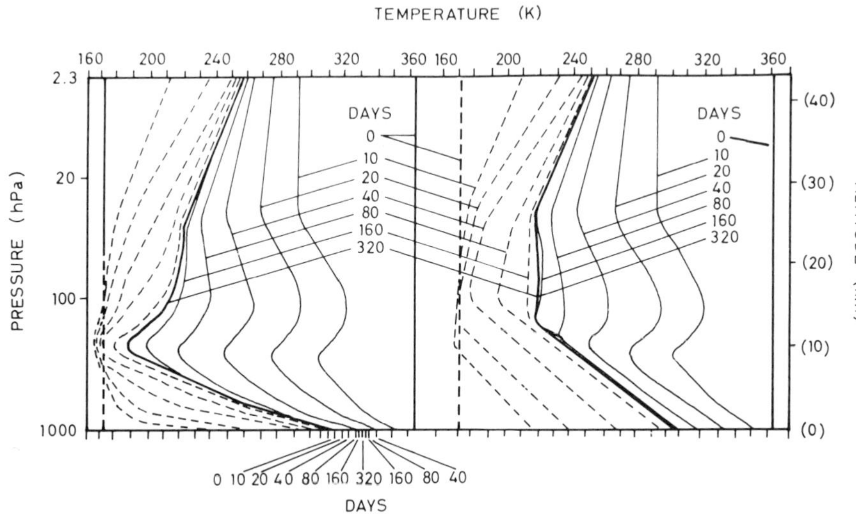

Figure 4.10 The left- and right-hand sides of the figure, respectively, show the approach to states of pure radiative and RC equilibrium. The solid and dashed lines show the approach from a warm and cold isothermal atmosphere (reproduced by permission of the American Meteorological Society from Manabe and Strickler, (1964), *Journal of the Atmospheric Sciences*, **21**, 361–385).

The complete structure of a basic RC model is shown in Figure 4.9. The flow continues until the atmospheric temperature profile converges to some final, equilibrium state. This convergence was well illustrated in one of the earliest works describing one-dimensional vertically resolved models (Figure 4.10). In this case two versions of the model are shown: convergence to purely radiative equilibrium and convergence to 'thermal' equilibrium. The latter is similar to the now more usual RC model except that in this early (1964) model the absolute humidity, rather than the relative humidity, was prescribed. The latter condition has since been shown to be more appropriate as, in reality, it varies little with atmospheric temperature. Thus the flow diagram in Figure 4.9 shows the calculation of the water vapour mixing ratio in atmospheric layers using the assumed, constant vertical profile of relative humidity.

4.5 SENSITIVITY EXPERIMENTS WITH RADIATIVE–CONVECTIVE MODELS

The RC model can be summarized by saying that the vertical temperature profile of the atmosphere plus surface system, expressed as a vertical temperature set, T_i, is calculated in a timestepping procedure, such that

$$T_i(z,\ t+\Delta t) = T_i(z,\ t) + \frac{\Delta t}{c_p \varrho}\left[\frac{dF_r}{dz} + \frac{dF_c}{dz}\right] \qquad (4.23)$$

Here the temperature, T_i, of a given layer, i, with height z and at time $t+\Delta t$ is a function of the temperature of that layer at the previous time t and the combined effects of the net radiative and 'convective' energy fluxes deposited at height z. In equation (4.23), c_p is the heat capacity at constant pressure, ϱ is the atmospheric density and dF_r/dz and dF_c/dz are the net radiative and convective flux divergences. There are two methods of implementing RC models: either the timestepping continues until there is a balance between the top-of-the-atmosphere shortwave and longwave fluxes, or the radiative flux changes immediately following a perturbation are computed. It was the latter method that was used to compute the impact of the eruption of Mount Agung on surface temperatures (Figure 1.13).

Radiative–convective models are particularly useful for studying the probable effects of perturbations to the radiative characteristics of the atmosphere. Their disadvantage is that they are formulated in terms of a global averaged value. On the other hand, this makes them computationally very efficient and more time can therefore be taken in making spectrally detailed calculations.

Sensitivity to humidity

Table 4.2 compares the predictions of ΔT (the increase in surface temperature) for differently formulated one-dimensional RC models for a perturbation in the form of a doubling of the atmospheric carbon dioxide from 300 to 600 ppmv. Model 1 has fixed absolute humidity. Hence the amount of water vapour in the atmosphere does not change and, in response to the external perturbation of CO_2, temperatures increase and relative humidity decreases. The resulting temperature increase of ~ 1.2 K can be thought of as a basic RC result since this model does not incorporate any feedback effects (i.e. neither atmospheric

Table 4.2 Equilibrium surface temperature increase due to doubled CO_2 (300–600 ppmv): results from a suite of one-dimensional model sensitivity experiments (modified with permission from Hansen *et al.*, 1981).

Model[a]	Description	ΔT (K)	Feedback factors[b] f	λ_{TOTAL}
1.	Fixed absolute humidity, 6.5 K km^{-1}, fixed cloud altitude.	1.22	1	3.75
2.	Fixed relative humidity, 6.5 K km^{-1}, fixed cloud altitude.	1.94	1.6	2.34
3.	Same as 2, except moist adiabatic lapse rate replaces 6.5 K km^{-1}.	1.37	0.7	5.36
4.	Same as 2, except fixed cloud temperature replaces fixed cloud altitude.	2.78	1.4	2.68

[a] Model 1 has no feedbacks affecting the atmosphere's radiative properties.
[b] The feedback factors f (dimensionless) and λ_{TOTAL} (W m^{-2} K^{-1}) are those defined in Section 1.4.4 and are two commonly used methods of representing the effect of each added process on model sensitivity to doubled CO_2.

water vapour nor clouds have changed in response to the temperature change and λ_{TOTAL}, as defined in Section 1.4.4, is equal to λ_B). Model 2 has, by contrast, fixed relative humidity. This means that as the temperature increases the saturation vapour pressure increases and thus, because the relative humidity is the ratio of actual vapour pressure to saturated vapour pressure, the actual vapour pressure must also increase. This extra water vapour must be the result of surface evaporation. It introduces a positive feedback of $\lambda = -1.41$ W m^{-2} K^{-1} so that λ_{TOTAL} decreases and the surface temperature increase predicted is 1.94 K. The difference between the results from models 1 and 2 illustrates the effect of evaporation on radiative exchanges and its importance for any climate prediction model. Model 3 uses a convective adjustment to the saturated adiabatic lapse rate rather than the value of 6.5 K km^{-1} that is used in models 1

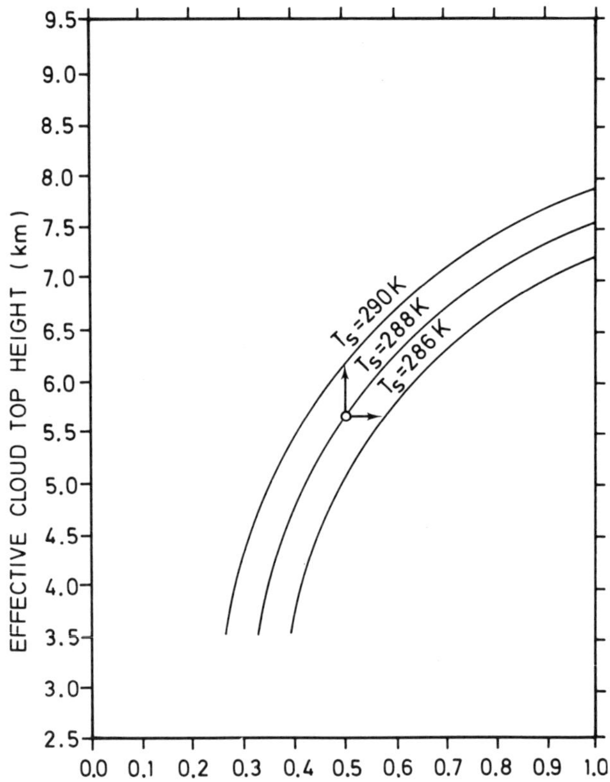

Figure 4.11 Effective cloud top height and cloud cover fraction giving the indicated equilibrium surface temperatures. From a present-day value of 0.5 for cloud cover and an effective cloud height of 5.5 km, increasing the cloud amount for constant cloud height decreases the surface temperature, whereas retaining the fractional cover at 0.5 and increasing the cloud height causes a surface temperature increase (reproduced with permission from Schneider, 1972).

and 2. It produces a slightly lower predicted temperature increase since the lapse rate decreases as additional water vapour is added to the atmosphere, i.e. two feedbacks of opposite sign combine. The difficulty of selecting an 'appropriate' global lapse rate to which convective adjustment should be made is considerable.

Comparison of models 4 and 2 illustrates the importance of cloud temperature and height effects. In model 4, the clouds, which are set at a constant, empirically determined amount, are at fixed temperatures rather than at the fixed heights used previously. The clouds therefore move to higher altitudes as the CO_2 perturbation increases temperature, affecting the resultant surface temperature. Clouds at a higher altitude radiate less energy to space, for a constant cloud cover, because their temperatures are lower. Hence the computed surface temperature must be raised further so that the planetary (top-of-the-atmosphere) energy balance is maintained, resulting in a predicted surface temperature increase, $\Delta T = 2.8$ K, which is considerably larger than the ~1.9 K for fixed cloud altitude.

Sensitivity to clouds

Absorption of solar energy by the entire Earth–atmosphere system is most simply specified in terms of the albedo of that system; usually using separate albedos for the cloudless part of the Earth–atmosphere system and the cloud-covered part. Using an RC model, it has been shown (Figure 4.11) that increasing the cloud amount by about 8 per cent (retaining fixed cloud top height and albedo) would lower the global mean surface temperature by 2 K, whereas raising the level of the cloud top height by about 0.5 km (at constant cloud amount and albedo) would produce exactly the opposite effect on surface temperature. Thus it is not possible to generalize globally averaged results to yield information on the local (in latitude and time) effect of variations in cloudiness, since the effect of changes in cloudiness on surface net heating depends upon the local values of the cloud amounts, heights and albedos, the albedo of the surface, the average solar zenith angle and the local vertical distribution of temperature and radiatively active constituents. The radiative calculations do suggest that even the direction of possible cloud feedback on surface temperature is far from obvious.

Sensitivity to lapse rate selected for convective adjustment

The generally accepted value of the critical lapse rate of 6.5 K km^{-1}, to which 'convective' adjustment is made, may not be appropriate for many experiments. Observational data give a globally averaged value for the tropospheric lapse rate closer to 5.5 K km^{-1}. Experiments have also been undertaken in which the moist adiabatic lapse rate rather than a fixed value was used. In one such model for a doubled CO_2 experiment the resulting change in surface temperature was 0.79 K as opposed to 1.94 K when the 6.5 K km^{-1} value was used. These differences are similar to those found in another, different, model in which a 2.36 K increase in surface temperature from a doubling of atmospheric CO_2 concentration was computed using a band-averaged calculation, whereas only a 1.9 K temperature increase was found when using a model which was identical except for the subdivision into many limited spectral intervals and where the CO_2–water vapour overlap region around 15 μm was divided into four sub-intervals. This lapse rate dependent difference is larger than that indicated in Table 4.2 where the temperature increase computed for doubled CO_2 using the saturated adiabatic lapse rate was 1.37 K.

4.6 DEVELOPMENT OF RADIATIVE–CONVECTIVE MODELS

Inclusion of an interactive cloud prediction scheme into a one-dimensional RC model would imply the incorporation of another facet of the climate pyramid (Figure 2.1), viz. dynamics.

There are a number of ways in which cloud–climate feedbacks can be investigated using one-dimensional RC models. Here we describe two developments of one-dimensional RC models which include a cloud prediction scheme.

4.6.1 Cloud amount and height predicted from 'convection'

The model used is a one-dimensional, globally averaged RC model in which the atmosphere is divided into seventeen layers. The tropospheric lapse rate is set at the 1962 US standard atmospheric lapse rate of 6.5 K km^{-1} and a fixed relative humidity is maintained.

Interactive cloud routine

The clouds are expressed as fractional parts of total cloud cover where C_i is the fractional cloud cover in layer i. Since overlapping clouds are not allowed, the total cloud cover, ΣC_i, will always be less than or equal to unity.

The cloud cover is predicted by being calculated as a proportion of the water mixing ratio, W, in each layer. This mixing ratio will be affected in turn by the latent heating, an indication of the amount of water added, and by precipitation, the water removed. The latent heat flux, H_L, will be determined from the model's convective adjustment and a calculated effective "atmospheric Bowen ratio", B. It is necessary to make an assumption about the variation of B with altitude. The simplest assumption is that B is constant, i.e. independent of altitude. There are data which seem to substantiate this assumption but it remains a somewhat dubious claim as the data do not take into account small scale turbulence.

The net latent heat flux into each layer is found from the convective adjustment which the model calculates at each timestep. The net convective flux into a layer, H_C, is then derived. This net total convective flux can be divided by $(1+B)$ to give the net latent heat flux

$$\frac{H_C}{1+B} = \frac{H_C}{1+(H_S/H_L)} = \frac{H_C \times H_L}{H_L + H_S} = H_C \frac{H_L}{H_C} = H_L \tag{4.24}$$

since H_C, the total convective flux, equals the sum of the latent flux and the sensible flux. These latent fluxes in each layer are then multiplied by the gravitational acceleration divided by the pressure change across the layer to give the flux per unit mass for unit area.

The cloud cover of the previous timestep multiplied by a constant gives the old water mixing ratio, W_o. This constant represents a typical mixing ratio for precipitating cloud systems. Thus greater cloud is linearly equated with higher mixing ratios and thus greater precipitation. Based on observational data it is reasonable to choose the value of the constant to be 5.5×10^{-4}.

The precipitation rate is calculated by multiplying the old mixing ratio, W_o, by another constant, representing the inverse of the conversion time from cloud droplets into rain. Again from observational data a value of $1.25 \times 10^{-4}\,s^{-1}$ seems representative of this conversion constant.

Thus the new water mixing ratio is the sum of three terms and

$$W = W_o + \frac{H_C}{(1+B)L} - 1.25 \times 10^{-4} W_o \tag{4.25}$$

the first term being the old ratio, the second the amount of water released through condensation and the third that precipitated out, where L is the latent heat of evaporation.

The new fractional cloud cover for each of i layers is then easily determined from

$$C_i \equiv C \text{ (for layer } i) = W/(5.5 \times 10^{-4}) \tag{4.26}$$

Finally, an adjustment is made if the total cloud coverage is greater than unity i.e. if $\Sigma C_i > 1$. In this case a proportional amount of cloud is removed from each layer by converting it to rain.

Sensitivity of a predictive cloud model

This model has been used to investigate the role that changing cloud cover and height may have played in the evolution of the Earth's atmosphere. Figure 4.12 shows three possible evolutionary states of the climate system. Case 1 has a lowered solar luminosity appropriate to the early Pre-Cambrian ($\sim 4.0 \times 10^9$ years ago) of 80 per cent of present-day solar flux, higher atmospheric CO_2 (1650 ppmv) and a surface albedo appropriate to a near global ocean (0.05). Despite the low value of incident solar flux, the computed surface temperature, 277 K, is well above freezing. Other situations can also be examined. If a greater emergence of land is postulated (case 2) then the surface albedo would be expected to be higher (say 0.10) and the computed surface temperature lower, 274 K. If, additionally, considerable silicate weathering is believed to have

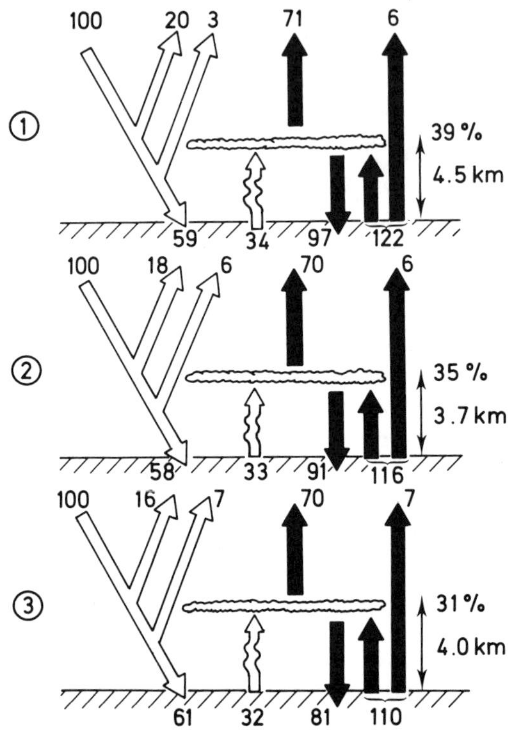

Figure 4.12 Three possible configurations of the atmosphere–hydrosphere of the early Earth. The calculated fluxes have been normalized to the incident solar radiation of the standard case 1 (i.e. 1100 units $= (1/4) \times 1100$ W m^{-2}). Solar fluxes (open arrows) are on the left of each sketch, upward arrows represent atmospheric (left) and surface (right) reflection. The wavy arrow represents convective heat flux from the surface and infrared fluxes are on the right (solid arrows). Note the variations in mean cloud cover (as a percentage) and cloud height (in kilometres) at right (reproduced with permission from Cogley and Henderson-Sellers, *Rev. Geophys. Space Phys.*, **22**, 161 (1984), Copyright by the American Geophysical Union).

occurred reducing the atmospheric CO_2 to below present-day values (say 330 ppmv) then the mean global surface temperature drops still further to 270 K.

It is important to realize that the inclusion of a cloud prediction scheme has caused this RC model to predict surface temperatures for these postulated early Earth situations which are significantly higher than would have been the case if clouds had been prescribed. In particular the decreasing cloud amount as temperatures fall and the relative increase in cirrus cloud, with its associated greenhouse effect, gives rise to surface temperatures at which sea-ice just melts in the present-day ocean. Thus results from this RC model which includes prediction of cloud amounts and heights (and hence properties) offer a possible solution to the 'reduced solar luminosity–enhanced early surface temperature' paradox described in Section 3.3.

4.6.2 A water vapour transport model

The model to be presented here couples water vapour transport interactively to an RC model. As the temperature profile changes in response to radiative heating or cooling, or latent heating, the distribution of water vapour is allowed to change in response to changes in the large scale vertical velocity field and eddy diffusion. Clouds are formed whenever the atmosphere is saturated and the latent heat is released to the atmosphere. Although the total cloud cover is fixed at 50 per cent, the number of cloud layers, their altitudes and thicknesses are computed. In addition, separate distributions of water vapour are computed for clear and cloudy conditions.

The distribution of water vapour in the troposphere is to a large extent controlled by dynamics. In most RC models water vapour is described by a constant distribution of relative humidity, $\psi(p)$, which differs from model to model but is typically of the form

$$\psi(p)=\psi(p_0)\left[\frac{p/p_0-b}{1-b}\right]^{\alpha} \tag{4.27}$$

where b, and α are constants, p is pressure and p_0 the surface pressure. The surface value of relative humidity, $\psi(p_0)$, is typically between 0.75 and 0.77.

A water vapour distribution given by equation (4.27) usually underestimates the amount of water vapour in the atmosphere. The annual average precipitable water described by the constant distribution of relative humidity is about 1.7 cm, 35 per cent less than the observed value of 2.6 cm. As a result, the greenhouse effect is underestimated.

In an atmosphere with clouds, a distribution of relative humidity will not be very well approximated by equation (4.27). Throughout a cloud the relative humidity is 100 per cent. Below the cloud the relative humidity generally decreases to the surface value. Therefore, it may be preferable to use separate distributions of relative humidity to model average conditions, one for a clear sky and one for a cloudy sky. The second one-dimensional RC cloud prediction scheme described here is based on such a representation.

For clear sky conditions, a constant distribution of relative humidity given by equation (4.27) with $b=0.02$ and $\alpha=1$ is used. Whenever $\psi(p)$ is less than zero, the mixing ratio is set at 3×10^{-6} kg kg^{-1}. For cloudy conditions, a water vapour transport model is used that includes large and small scale vertical velocities. When the computed water vapour density exceeds the saturation value, a cloud is assumed to be formed. A separate transport equation describes the liquid water (or ice) content in the cloud region. The method predicts cloud height and thickness, the water vapour distribution above and below the cloud and the distribution of condensed material within the cloud. Latent heat released is returned to the atmosphere. Multiple cloud layers (up to three) can be handled in the radiation model. This model, in contrast to the model described in Section 4.6.1, does not predict changes in cloud amount which is assumed to be 50 per cent.

Transport of water vapour is given by

$$\frac{\partial^2 q}{\partial z^2} = \left[\frac{w}{K_z} + \frac{1}{H_\varrho(z)} \right] \frac{\partial q}{\partial z} \tag{4.28}$$

where the second term on the right-hand side is a source/sink term and q is the water vapour mixing ratio, w, the large scale vertical velocity, K_z, a mean eddy exchange coefficient, and $H_\varrho(z)$ the density scale height. The large scale vertical velocity is assumed to be the velocity necessary to maintain hydrostatic equilibrium. This calculation is based on the Richardson equation in which all driving forces (i.e. pressure gradients, velocity divergences, heating terms, etc.) at a specific altitude are assumed to have a magnitude equal to the non-radiative heating rates necessary to maintain energy balance. The average vertical velocity produced under this assumption is of the order of 0.003–0.006 m s^{-1}. The model calculations are very insensitive to the individual values of w and K_z and so average atmospheric values are used. Whenever the computed value of the mixing ratio exceeds the saturation value, the mixing ratio is assigned the saturation value.

In this way, the region of the atmosphere where a cloud should exist can be determined. Equation (4.28) is solved numerically. The mixing ratio at the surface, $q(p_0)$, is given by the surface relative humidity $\psi(p_0)$ and the surface temperature. The 'top' of the model, or the extent over which water vapour is transported vertically, is prescribed. This assumed pressure level can be compared to the level of zero divergence in the atmosphere (~500 hPa). The latent heat released following condensation or deposition is returned to the atmosphere and the temperature profile inside the clouds adjusted accordingly.

With the exception of the terminal velocity of the droplets, all of the relevant variables are coupled to the temperature profile. The value of the water vapour mixing ratio at the surface is set equal to that for a relative humidity of 77 per cent, in accordance with the widely used constant relative humidity assumption. While the model can generate multiple clouds, typically only one cloud is found in the equilibrium atmosphere.

The cloud and radiation models are coupled by first generating a set of clear sky heating/cooling rates which are used to determine w. The water vapour transport model gives the water vapour distribution and cloud heights. These are then used in the radiation program which computes a temperature profile in RC equilibrium with flux balance at the top of the atmosphere. The calculations are then repeated until the water vapour, clouds and temperature are self-consistent. The water vapour calculations are made at 37 equal height increments from the surface to the model top. These results are then interpolated on to the coarser grid used in the radiation routine.

The equilibrium water vapour mixing ratio profile is shown in Figure 4.13 for a 35 °N, April atmosphere. The calculations were made with a model top of 450 hPa and resulted in a thin cloud centred at 500 hPa. Also shown is the profile

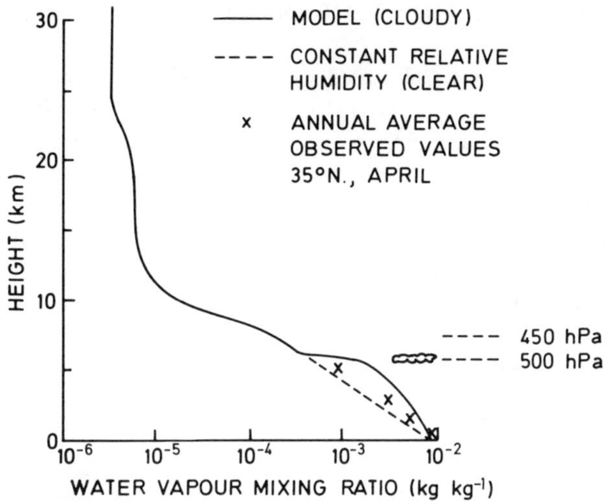

Figure 4.13 Comparison of the results of a water vapour transport RC model using different lapse rate schemes as compared with observations (reproduced with permission from Hummel and Kuhn, 1981).

resulting from the constant relative humidity assumption used for the clear sky distribution. The average of the clear sky water vapour distribution and the cloudy sky result (with a model top at 450 hPa) yielded the best agreement with the zonally averaged annual observations, and the equilibrium temperature profile from the coupled radiation–transport model for this atmosphere gives good agreement with standard atmosphere profiles. In addition, the equilibrium cloud is found at a height of 5.52 km in agreement with the constant cloud top height of 5.5 km employed in many non-predictive RC models.

Model sensitivity

In this model the radiation field determines the temperature when saturation does not occur; otherwise the moist adiabatic lapse rate is used. The cloud coverage is fixed at 50 per cent and there are separate calculations for the clear and cloudy sky, the radiation fluxes then being weighted by the cloud cover to yield the final temperature.

The temperature increase for a doubling of CO_2 from this model is 0.8 K. This value is considerably lower than the temperature increase predicted by most GCMs and is less than all the one-dimensional RC predictions listed in Table 4.2.

Clearly, there is still much improvement to be made in one-dimensional models of all types and, although some modellers have succeeded in coupling a 'cloud generation' model with a detailed radiation code, the cloud cover or other attributes must still be specified. These difficulties, and many others, also beset the general circulation climate modellers as will be described in Chapter 6.

4.7 RADIATION: THE DRIVER OF CLIMATE

In this chapter we have emphasized radiation because it is fundamental to the climate. Solar radiation is absorbed and infrared radiation emitted, with these two terms balancing over the globe when averaged over a few years. This simple energy balance was the basis of the EBMs described in Chapter 3, but then the way in which radiation is absorbed, transferred and re-emitted was ignored. In this chapter we have stressed the radiative transfer processes: the heating of the surface by absorption of shortwave energy and the heating and cooling of the atmosphere by absorption and emission of infrared radiation.

The radiative equilibrium attained is usually unstable and so we were forced to couple a convective adjustment into our radiatively driven model. These simple radiative–convective responses were the basis of the first descriptions of the ''nuclear winter'' phenomenon. Large fires following a nuclear exchange cause the input of large amounts of high elevation aerosols. These carbon particulates absorb solar radiation and heat the upper parts of the troposphere thus dramatically reducing solar input to the surface *and* causing the atmospheric lapse rate to become strongly stable. Convection ceases and the cold surface becomes decoupled from the atmosphere. Recovery from this climatic configuration requires a large input of energy.

The convective adjustment which usually occurs in the atmosphere permits the introduction of vertical motion into our climate model and can use this to parameterize movement of atmospheric water and the formation of clouds. The approximations we have made are acceptable for global modelling, but once we become interested in a specific region we must consider horizontal transfer as well as vertical motion and radiation. In Chapter 5 we introduce the concept of large scale horizontal motion.

RECOMMENDED READING

Chylek, P., and Kiehl, J. T. (1981) Sensitivities of radiative-convective models, *J. Atmos. Sci.*, **38**, 1105–1110.

Hansen, J. E., Johnson, D., Lacis, A. A., Lebedeff, S., Lee, P., Rind, D., and Russell, G. (1981) Climate impact of increasing atmospheric CO_2, *Science*, **213**, 957–966

Hummel, J. R., and Kuhn, W. B. (1981) An atmospheric radiative-convective model with interactive water vapour transport and cloud development, *Tellus*, **33**, 372–381.

Manabe, S., and Wetherald, R. T. (1967) Thermal equilibrium of the atmosphere with a given distribution of relative humidity, *J. Atmos. Sci.*, **24**, 241–259.

Ramanathan, V., and Coakley, J. A. (1979) Climate modelling through radiative-convective models, *Rev. Geophys. Space Phys.*, **16**, 465–489.

Rossow, W. B., Henderson-Sellers, A., and Weinreich, S. K. (1982) Cloud-feedback — a stabilizing effect for the early Earth, *Science*, **217**, 1245–1247.

Schneider, S. H. (1972) Cloudiness as a global climatic feedback mechanism: the effects on the radiation balance and surface temperature of variations in cloudiness, *J. Atmos. Sci.*, **29**, 1413–1422.

Wang, W-C., and Stone, P. H. (1980) Effects of ice-albedo feedback on global sensitivity in a one dimensional radiative-convective model, *J. Atmos. Sci.*, **37**, 545–552.

Chapter 5

Two-Dimensional Models

'It is not clear whether the authors [*of a previous paper*] intended to question the necessity of meridional circulations for the maintenance of the kinetic energy of the atmosphere. Such an intention would mean a complete change of the whole foundation of dynamic meteorology, and I doubt strongly that the genius of the authors recognized by all meteorologists, will be sufficient for that goal.'
E. Palmén (1949) (*Journal of Meteorology*, **6**, 429–430)

5.1 WHY TWO-DIMENSIONAL MODELS?

There is an excellent book by Abbott which describes a world called 'Flatland' which is inhabited by two-dimensional beings and, finally, visited by a strange three-dimensional object: a sphere. The sphere passes through Flatland and is perceived by the inhabitants as being only a series of discs of changing radius. This glimpse of the three-dimensional 'reality' is impossible for most Flatlanders to comprehend. Climate modellers, on the other hand, are only too painfully aware of the multi-dimensional nature of the climate system. Those who design and work with two-dimensional models are not uncomprehending of the missing dimension but have chosen to use a simpler model type. There are two reasons: (i) two-dimensional models are simpler and therefore cheaper to integrate on computers and thus can be used for much longer timescale integrations than full GCMs and (ii) being simpler, two-dimensional models therefore represent particular features of the climate system more simply because other confusing features are removed. Thus modellers, unlike Flatlanders, recognize complexity and intentionally seek to reduce it. In this chapter we explore some of their reasons and results.

5.2 TWO-DIMENSIONAL STATISTICAL DYNAMICAL CLIMATE MODELS

Secondary school Geography classes would appeal to Flatlanders because at this level the general circulation of the atmosphere is often introduced as being composed of cellular circulations. These Hadley, Ferrel and polar cells are meridional features, i.e. they consist solely of latitudinally averaged movement between zones (Figure 5.1). Of course this is a gross simplification which ignores

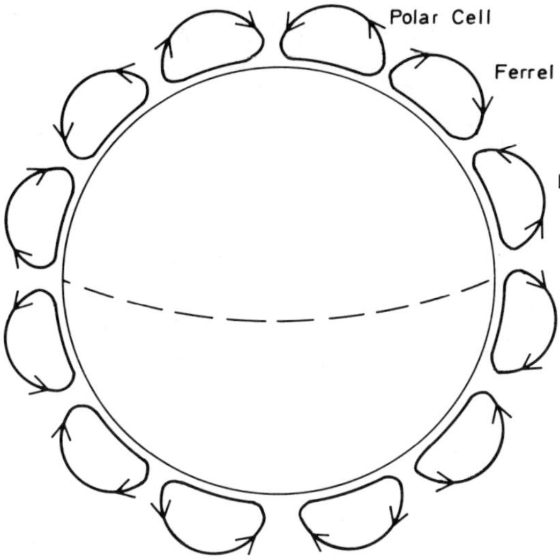

Figure 5.1 Dominant features of the meridional circulation (stylized). The three cells are illustrated, the strongest of which is named the Hadley cell after George Hadley (1685–1768) who formulated a theory of the trade winds in 1735 but which went unnoticed until 1793. (Hadley was also the first to recognize the importance of what later came to be known as the Coriolis force).

the major circulation features in mid-latitudes: the Rossby waves (see Figure A.1 in the glossary in Appendix A). Most two-dimensional SD climate models are constructed to simulate the meridional motions only. The two dimensions they represent explicitly are height in the atmosphere and latitude. Variations around latitude zones (i.e. longitudinal variations) are neither resolved nor described. These models solve numerically the basic equations listed in Table 2.1. In general these models are able to produce realistic simulations of the large scale two-dimensional flow (Figure 5.2). The fundamental difference between these models and full atmospheric GCMs is that all the variables of interest are zonally averaged values. This zonal averaging is identified below by angle brackets. The equations to be solved are:

Zonal momentum

$$\frac{\partial <u>}{\partial t} - f<v> + \frac{\partial <u'v'>}{\partial y} = F \tag{5.1}$$

Meridional momentum (geostrophic balance)

$$f<u> + R<T> \frac{\partial}{\partial y} (\ln <p>) = 0 \tag{5.2}$$

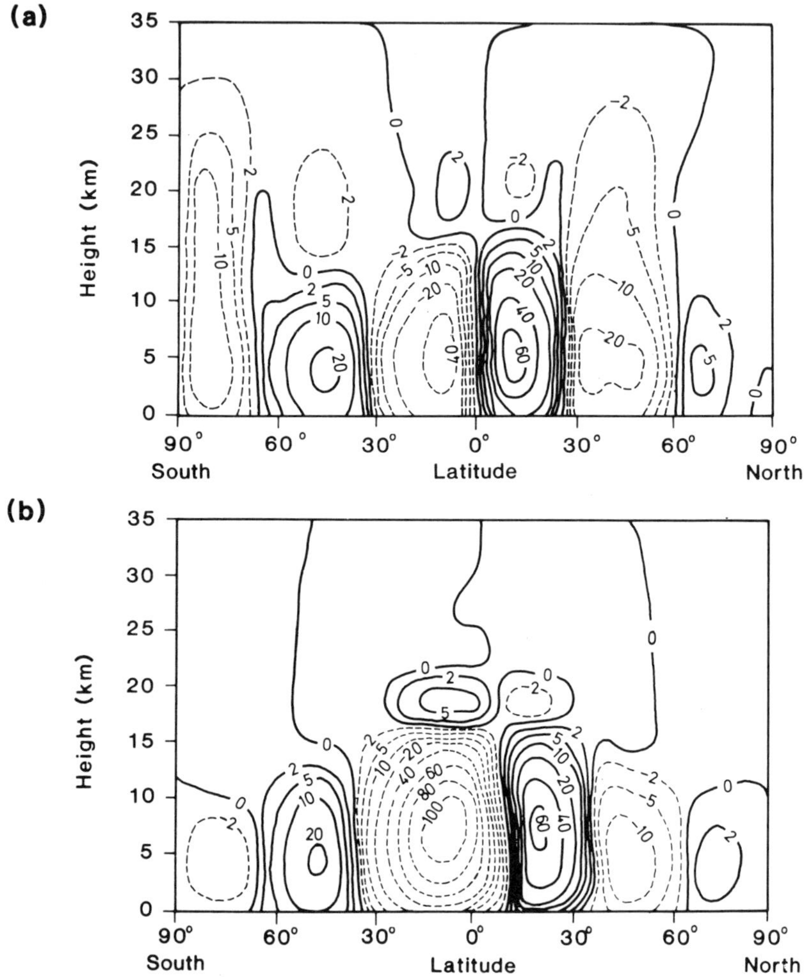

Figure 5.2 Annual and zonal mean meridional mass flux stream function (10^9 kg s^{-1}): (a) as observed (Oort and Peixóto, 1983); and (b) as simulated by the Lawrence Livermore two-dimensional SD model (reproduced with permission from MacCracken and Ghan, 1987).

Hydrostatic balance (vertical component)

$$\frac{\partial(\ln <p>)}{\partial z} = -\frac{g}{R<T>} \tag{5.3}$$

Thermodynamic balance

$$\frac{\partial <T>}{\partial t} + \frac{\partial <v'T'>}{\partial y} + \frac{\partial <w'T'>}{\partial z} + \\ <w> \left\{ \frac{g}{<\varrho> c_p} + \frac{\partial <T>}{\partial z} \right\} = \frac{Q}{<\varrho> c_p} \tag{5.4}$$

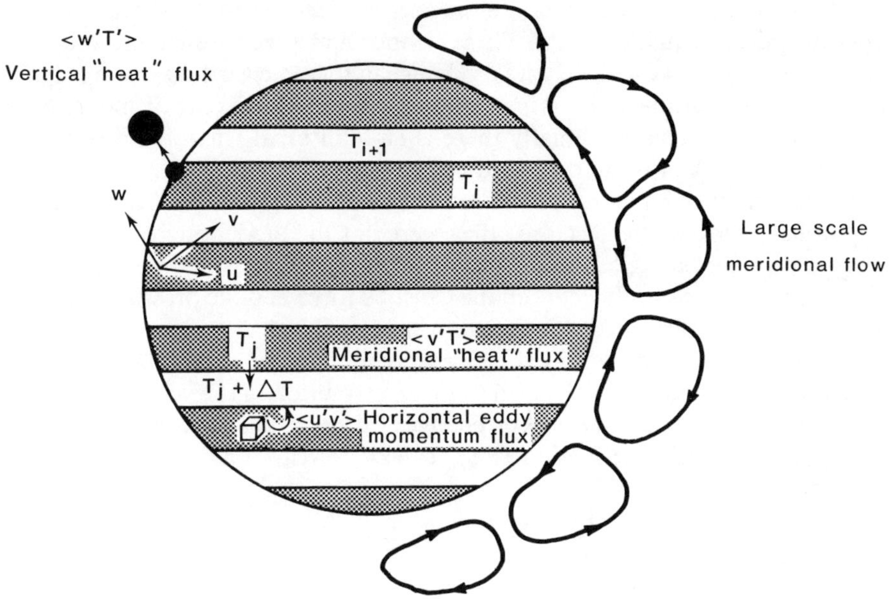

Figure 5.3 Schematic diagram of the fluxes and variables represented in a two-dimensional SD model. The "heat" fluxes require premultiplying by ϱc_p to be complete and the eddy momentum flux requires premultiplying by ϱ.

Continuity

$$\frac{\partial <\varrho><v>}{\partial y} + \frac{\partial <\varrho><w>}{\partial z} = 0 \qquad (5.5)$$

where u, v and w = the velocities in the eastward (x), northward (y) and vertical (z) directions,
T = the temperature,
Q = the zonal diabatic heating,
F = friction term,
R = universal gas constant,
c_p = specific heat at constant pressure,
g = acceleration due to gravity,
f = Coriolis parameter and
$<\varrho> = <p>/R<T>$.

The primed notation denotes a deviation from the zonal average of these variables.

As these equations are essentially those which are solved in GCMs, although here they are written in a simpler form, they are worth considering in some detail. The momentum equations are themselves expressed for unit mass, so that the density terms cancel and the momentum (mass times velocity) is represented simply by the velocity component in the direction of interest. Zonal momentum changes (with time) are thus represented by the first term on the left-hand side of equation (5.1). These temporal changes are balanced by the Coriolis term, $f\langle v\rangle$, and the rate of change in the poleward direction (Figure 5.3) of the correlation term, $\langle u'v'\rangle$, i.e. the eddy transport of momentum in the poleward direction. Finally there is an additional frictional dissipation term, F, to be taken into consideration.

The meridional momentum equation is similarly constructed but the small temporal changes are neglected; if they were not the model might accumulate errors and predict non-zero momentum fluxes at the poles. Consequently the balance equation is simply between the Coriolis force and the pressure gradient force in the poleward direction, friction being neglected.

The hydrostatic equation is the third component of the conservation of momentum equation (Table 2.1). In this case resolution in the vertical direction yields no Coriolis component, but changes occur in the pressure field (with height). This equation (equation (5.3)) can be rewritten as

$$\frac{1}{\langle p\rangle}\frac{\partial\langle p\rangle}{\partial z} = -\frac{g}{R\langle T\rangle} \tag{5.6}$$

whence, from the ideal gas law, is derived the more common expression:

$$\frac{\partial\langle p\rangle}{\partial z} = -\frac{g\langle\varrho\rangle R\langle T\rangle}{R\langle T\rangle} = -g\langle\varrho\rangle \tag{5.7}$$

The thermodynamic balance exists between the temporal rate of change of zonally averaged temperature and the rate at which temperature is both transported into and out of each latitude zone. This is accomplished by eddies in the lateral (northward) and vertical directions and represented by the two eddy correlation terms (second and third terms on the left-hand side of equation (5.4)). A further term represents vertical transport, taking into account adiabatic heating and cooling due to the compressibility of the atmosphere. The balance is completed by the inclusion of the diabatic heating term on the right-hand side.

The continuity equation says simply that mass can neither be created nor destroyed, i.e. the rate of change of mass in all three dimensions overall is zero. However, since zonal averages are under discussion, the change in the x direction has been averaged out, as expressed by the use of angle brackets, and only two components remain. The sum of these two is zero. Thus in regions where there is net divergence or convergence, there must of necessity be a vertical motion also.

In writing and discussing these ‘prototype’ equations we have neglected the need for a spherical geometry for a complete treatment. Despite this, the representation is useful. The horizontal component of the eddy momentum flux $\langle u'v'\rangle$ is not only responsible for transferring zonal momentum but also drives the meridional circulations shown in Figure 5.1. Figure 5.4 shows mean meridional cross-sections of (a) zonal wind, (b) meridional wind, (c) vertical motion and (d) temperature. These can be calculated by specifying a vertically

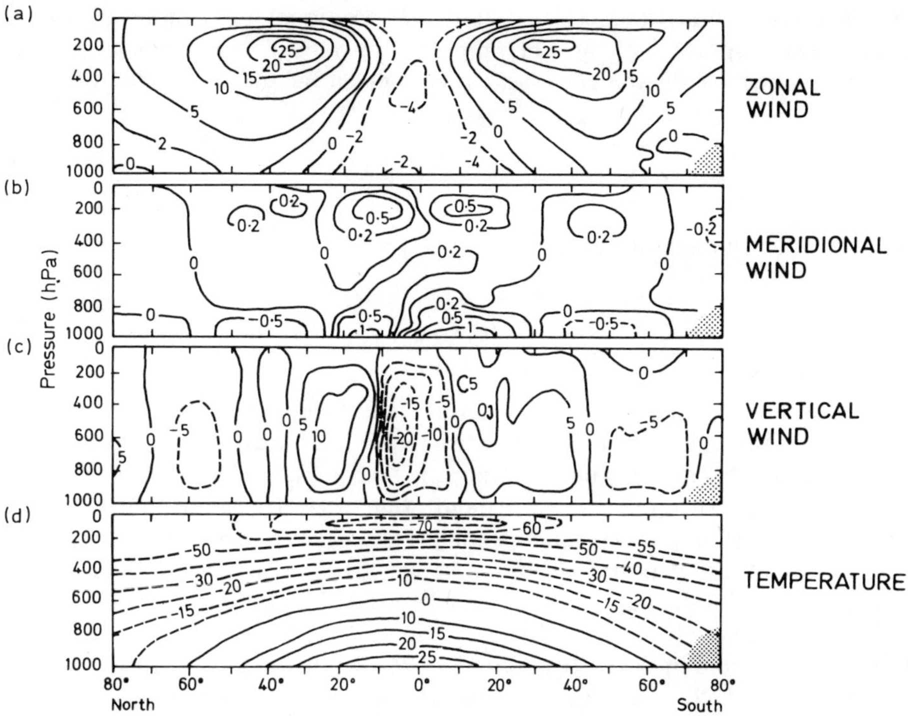

Figure 5.4 Observed zonal statistics for four atmospheric variables: (a) zonal wind ($m\,s^{-1}$); (b) meridional wind ($m\,s^{-1}$); (c) vertical wind ($\times 10^{-3}\,m\,s^{-1}$); and (d) temperature (°C) (reproduced by permission of Academic Press from Oort and Peixóto, 1983).

and latitudinally varying distribution of eddy momentum fluxes similar to those observed. Note that the magnitude of the three induced velocities decreases considerably: the zonal wind can be as high as $30\,m\,s^{-1}$ whilst the meridional wind is typically $0.25\,m\,s^{-1}$ and the induced vertical motion is $0.005\,m\,s^{-1}$.

For two-dimensional climate models it is necessary to find representations for the eddy fluxes in equations (5.1)–(5.5) so that this system of equations can be solved numerically. In these models the atmosphere is represented on a latitude versus pressure (height) grid with up to approximately eleven layers and ten to twenty grid points between the poles. Often considerable effort goes into representing atmospheric radiative processes and surface features although the main problem remains the characterization of eddy transports.

Eddy transport is of critical importance for determining the equator to pole temperature gradient and the vertically distributed zonal wind field, especially the strength of the jet-stream winds. Early parameterizations of eddy flux simply related eddy transports to gradients of zonal mean variables using empirically determined diffusion coefficients. This representation is quite similar to the parameterization used in some EBMs for the meridional energy transport (e.g. equation (3.12) in Section 3.2). This parameterization was based on the argument that, since baroclinic waves are driven by the meridional temperature gradient, their eddy transports might also be simply parameterized as being proportional to this gradient. Thus the eddy heat flux is given by

$$<v'T'> = -K_T \frac{\partial <T>}{\partial y} \tag{5.8}$$

and the eddy momentum flux by

$$<u'v'> = -K_M \frac{\partial <u>}{\partial y} \tag{5.9}$$

where K_T and K_M are empirically derived coefficients for temperature and momentum. More detailed study has shown that, while the diffusive representation is fairly reasonable for eddy heat transport, it is a completely inadequate representation of eddy momentum flux, since momentum can be transported up as well as down the meridional gradient of momentum. Consequently, later parameterizations reformulated the transport equations in terms of the potential vorticity gradient.

Up to this point the parameterizations described for two-dimensional models have been empirically based. However, subsequent theoretical analysis by a number of authors in the early and mid-1970s demonstrated that the diffusion coefficients as well as the eddy transport itself may be proportional to the meridional temperature gradient. It was found that the equator to pole temperature gradient was considerably different when computed with EBMs which used a value for the diffusion coefficient dependent on the temperature

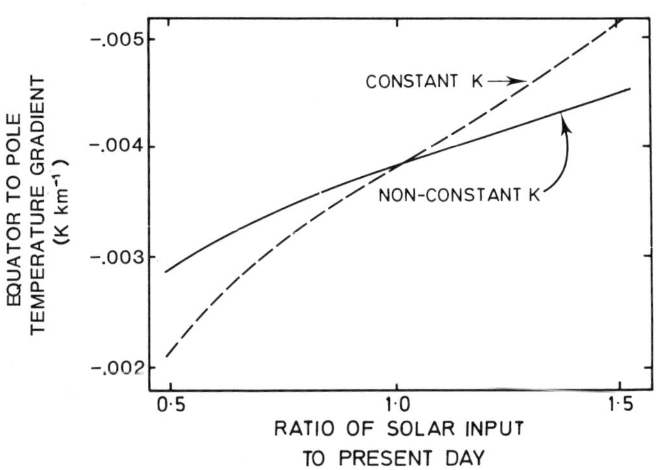

Figure 5.5 The effect of the inclusion of a variable (i.e. itself a function of the temperature gradient) diffusion coefficient, K, on the equator-to-pole temperature gradient in an EBM (reproduced by permission of the American Meteorological Society from Stone (1973) *Journal of the Atmospheric Sciences*, **30**, 521–529).

gradient as opposed to a constant eddy diffusion coefficient (Figure 5.5). This finding suggests that one of the reasons why the energy balance climate models (EBMs) showed considerable sensitivity to a small decrease in solar constant was that the diffusion coefficient in these models was constant, rather than being a function of the temperature gradient. A much greater decrease in the solar constant is necessary to initiate an ice age in a model which includes a temperature gradient dependency of the diffusion coefficient. This is because the temperature gradient remains high at low values of solar input. This recognition may offer another partial solution to the cool Sun–enhanced early surface temperature paradox described earlier (see especially Sections 3.3 and 4.6.1).

The basis of the parameterization problem is the simplification which is generally made in the solution of the zonal flow equation in baroclinic wave theory. The usual simplifications are to assume that the zonal wind, $<u(y, z)>$, is a function of y only (the barotropic solution) or z only (the baroclinic solution). Instability to small disturbances of a zonal wind which varies only in the vertical (z) direction (a baroclinic instability) converts to eddy energy the energy that is stored in the current latitudinal variation of zonal temperature, $<T>$. This energy is released by the eddy flux of heat $<v'T'>$. On the other hand, instability of a zonal wind to horizontal (y component) perturbations (a barotropic instability) converts kinetic energy of the zonal wind to eddy energy through the flux of horizontal eddy momentum $<u'v'>$.

The parameterization of momentum fluxes is considerably more complex when theoretically based than the parameterization of heat fluxes. It has been shown that potential vorticity is more suitable for the treatment of $<u'v'>$ as the eddy momentum flux can be obtained once eddy potential vorticity and eddy heat fluxes have been derived. A further problem with eddy flux parameterizations is the existence in the atmosphere of large stationary eddies forced both by topography and by land/ocean temperature contrasts. Since the parameterization schemes described above represent only transient baroclinic eddies, it is possible that there will be an underestimation of the total eddy transport produced due to the neglect of these stationary eddies. However, observational data suggest that a compensatory mechanism may exist since total eddy flux seems to be correlated better with observed temperature gradients than is the transient eddy flux alone.

Full formulations of two-dimensional SD models often also include vertical and horizontal eddy transports of water vapour as well as those of heat and momentum described above. Since two-dimensional models attempt to parameterize only the eddy transport, while the mean meridional transport terms are computed explicitly, it is hoped that inadequacies in the eddy transport parameterizations are compensated for in the explicit calculation of meridional transport.

5.3 CONVECTION, CLOUD COVER AND PRECIPITATION IN TWO-DIMENSIONAL STATISTICAL DYNAMICAL MODELS

Vertical motion must be parameterized rather carefully in zonally averaged models. Around any one latitudinal band, there would normally be a range of different surface types and atmospheric states possessing different degrees of instability. The extent of these areas and of the instability would be expected to vary with season and time of day. Unfortunately zonally averaged conditions can be quite stable even though within a latitude zone there might be many locations where convection would occur. Thus if cloud cover and precipitation are to be computed realistically, an 'on–off' switch in a zonally averaged model would result in zones being either cloudy with precipitation or having clear skies and no rain.

Instead of using the simple convective adjustment, whereby unstable lapse rates are altered to be equal to a predetermined lapse rate (usually 6.5 K km^{-1}) as is done in one-dimensional RC models (Chapter 4), a more complex, and usually empirically derived, formulation is required to determine the onset and extent of convection and often of cloud formation processes. Typically, adjacent vertical layers are compared and a formula which depends upon thresholds of relative humidities as well as atmospheric stability is applied. Generally less than 100 per cent cloud cover is predicted within individual layers in an attempt, once again, to avoid 'on–off switching'. Recently similar sub-gridscale descriptions of vertical convection and cloud formation processes have been derived to limit the extent of precipitation and/or cloud within a grid cell of a GCM (see Chapter 6).

Convective precipitation is likely to be the major contributor to rainfall in zonally averaged models. However, large scale precipitation can also occur if the mixing ratio of an air parcel exceeds the saturation mixing ratio. Since it is most unlikely that zonally averaged relative humidities would ever reach saturation in anything other than the near surface layer, thresholds are often set somewhat lower, say for instance a zonally averaged relative humidity of 80 per cent. If this threshold is exceeded, water vapour condenses and latent heat is released until the maximum permissible value of 80 per cent relative humidity is re-established. Evaporation of rainfall may occur if it falls through atmospheric layers which are not saturated. Precipitation is treated as snow rather than rain if the surface temperature, or sometimes the surface air temperature, is below freezing at that timestep. Clouds are usually assumed to be randomly correlated in the vertical so that the total cloud fraction, C, for a given latitude band is

$$C = 1 - \prod_{i=1}^{n} (1 - c_i) \tag{5.10}$$

where c_i are the cloud amount fractions in layers 1 to n.

5.4 RADIATION AND SURFACE CHARACTERIZATION IN TWO-DIMENSIONAL STATISTICAL DYNAMICAL MODELS

5.4.1 Radiation

In general two-dimensional models pay as much attention to radiative fluxes as do one-dimensional RC models. The treatment is generally similar to that described in Chapter 4, in that the solar radiation calculation includes the effect of scattering by air molecules, clouds and aerosols and usually absorption by water vapour, ozone, and aerosol and cloud particles. The longwave thermal fluxes include absorption and reradiation by clouds, water vapour, carbon dioxide and usually other atmospheric constituents.

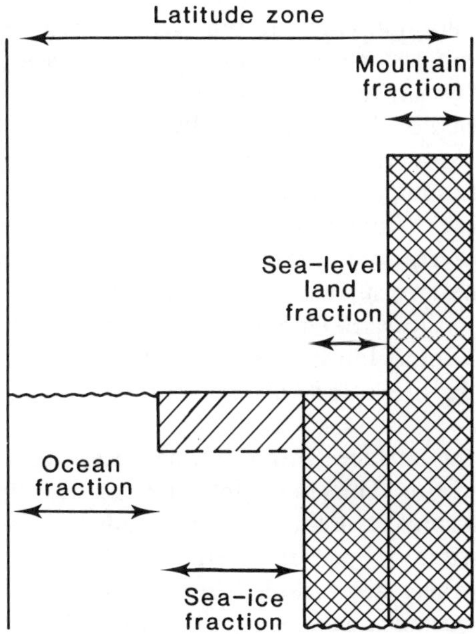

Figure 5.6 Conceptual view of fractional surface coverage used in calculating fluxes in the vertical and in the lowest atmospheric level in a sophisticated two-dimensional climate model. The proportions of each of the surface types can differ from zone to zone (after MacCracken *et al.*, 1981).

Solar insolation at the top of the atmosphere can be handled either by specifying a mean value which is held constant for the 24-hour day, or a diurnal variation can be imposed upon the computed seasonal variation. It has been shown that a seasonal variation in insolation is required if snow and ice are to build up and persist, since continuous lower levels of solar radiation fail to give rise to an adequate decrease in net surface heating to permit the establishment of these features. Similarly, convective effects are considerably suppressed if diurnally averaged irradiance is prescribed. Sometimes two or more versions of a model will coexist so that different features can be examined using different averaging and parameterization procedures.

As was described in Section 5.3, it is unusual for the simple convective adjustment, common in one-dimensional RC models, to be applied in two-dimensional models, since this would produce rather extreme 'switching' between convection/non-convection states for whole latitude zones.

5.4.2 Surface characterization

Two-dimensional SD models differ considerably in the degree to which the surface parameterization attempts to represent the real world. In the simplest case, mean zonal values for surface albedo and surface thermal capacity are prescribed in a fashion similar to that described for one-dimensional RC models except that, in the case of the latter, globally averaged values are required. The other extreme is illustrated in Figure 5.6, in which the zone is characterized as having variable fractional covers of ocean, sea-ice, sea-level land and higher land. Naturally these different components will all possess different thermal and radiative characteristics. For example, Table 5.1 lists the characteristics that are specified for the five surface types which are permitted in the Lawrence Livermore two-dimensional SD model. This is probably the most sophisticated two-dimensional SD model in its treatment of surface characteristics and has been used to study the effect of land-surface changes such as tropical deforestation.

Partitioning of the surface within a zone necessitates partitioning of the atmospheric column also. Thus in models with detailed surface characteristics, radiative fluxes and convective fluxes will probably be computed separately for each of the permissible or specified surface types.

Where ocean surfaces are included in two-dimensional models, it is also possible to specify oceanic heat fluxes so that part of the meridional transport of energy can be subsumed into this oceanic transport term. These models can also be used to perform more detailed studies of ice–albedo feedback since sea-ice and land snow cover can be computed independently within a latitude zone, and surface temperatures associated with the different surface types can be computed and stored independently of one another.

Surface to atmosphere fluxes of momentum, sensible and latent heats are computed using standard bulk aerodynamical formulae:

$$H_{\mathrm{D}} = \varrho c_D u^2 \tag{5.11}$$

$$H_{\mathrm{s}} = \varrho c_{\mathrm{p}} c_{\mathrm{s}} u (T_{\mathrm{s}} - T_{\mathrm{a}}) \tag{5.12}$$

$$H_{\mathrm{L}} = \varrho L c_{\mathrm{L}} u (q_{\mathrm{s}} - q_{\mathrm{a}}) \tag{5.13}$$

where c_D, c_{s} and c_{L} are the aerodynamic drag coefficients, u the wind speed evaluated at the same standard height as the drag coefficient, T the temperature, q is the water vapour mixing ratio and L the latent heat of vaporisation. This type of parameterization of surface to atmosphere fluxes is exactly the same

Table 5.1 Surface characteristics in the Lawrence Livermore two-dimensional SD model (after MacCracken *et al.*, 1981)

Surface type	Albedo	Density ($kg\ m^{-3}$)	Heat capacity ($J\ kg^{-1}\ K^{-1}$)	Thermal conductivity ($W\ m^{-1}\ K^{-1}$)	Atmosphere–surface interaction coefficient: Surface drag	Atmosphere–surface interaction coefficient: Thermal	Atmosphere–surface interaction coefficient: Evaporation	Runoff coefficient (fractional loss per day)
Ocean	Direct and diffuse values computed as a function of solar elevation	1015.0	4184.0	—	1.1×10^{-3}	1.8×10^{-3}	1.1×10^{-3}	—
Snow	0.7[a] for dry snow ($T<273.2K$)							
	0.55[a] for wet snow	400.0	878.6	1536.0	2.2×10^{-3}	2.2×10^{-3}	2.2×10^{-3}	0.50
Ice	0.70[a] for dry ice ($T<273.2K$)							
	0.55[a] for melting ice	917.0	2008.0	8054.0	2.2×10^{-3}	2.2×10^{-3}	2.2×10^{-3}	0.25
Tundra	0.15 if $\psi_s<0.25$[b]							
	0.30 if $\psi_s>0.70$	1600.0	836.8	8786.0	4.3×10^{-3}	4.3×10^{-3}	4.3×10^{-3}	0.25
Savannah	0.10 if $\psi_s<0.25$							
	0.25 if $\psi_s>0.70$	1600.0	836.8	8786.0	4.3×10^{-3}	4.3×10^{-3}	4.3×10^{-3}	0.25

[a] For snow and ice less than 0.04 m thick, the albedo of the underlying surface is used. For snow and ice between 0.04 and 0.1 m thick, the albedo of the underlying surface is based on a linearly increasing snow albedo from 0 to 0.1 m thickness.

[b] ψ_s is the surface humidity.

as is used in GCMs (see Chapter 6). Within two-dimensional models, however, it is necessary to parameterize quasi-longitudinal coupling between surface types if more than one type is permitted in a given latitude zone. Such parameterizations are semi-empirical and are often based on the concept of necessary response times for full equilibration; one such response time is associated with each surface type.

It will be clear by now that if this level of detailed parameterization is being sought it might well be preferable to utilize a fully three-dimensional model such as the general circulation climate models described in Chapter 6. However, two-dimensional zonal models are usually more economical of computer time than three-dimensional models and can therefore be applied to long time period climatic simulations more readily than can full GCMs.

The most useful area of application of two-dimensional SD models is probably predicting the effects of *small* perturbations. Good examples of current interest are the increase in Arctic aerosol loading and transient increases in stratospheric and tropospheric aerosol. Two-dimensional SD models are especially useful in such studies for two reasons: (i) because detecting the signal produced by such small changes in a full three-dimensional model with eddy-related inherent variability would take many years of climate simulation, and (ii) because the slight disturbances in the climatic state do not negate the assumption inherent in the formulation of two-dimensional models—that the eddy fluxes can be satisfactorily parameterized. A full two-dimensional SD model should be approximately 100 times faster in execution than a GCM of roughly equivalent resolution and physical detail.

5.5 INTERCOMPARISON OF A TWO- AND A THREE-DIMENSIONAL MODEL

Although three-dimensional models will not be introduced until the next chapter, a brief comparison of a relatively simple three-dimensional atmospheric GCM and a relatively sophisticated two-dimensional SD model is useful here. The comparison was undertaken by a pair of scientists, one drawn from each of the two modelling groups and so should be fair to both models.

The purpose of the comparison was to examine the performance of the two selected climate models under as closely comparable conditions as possible and to try to interpret the results in terms of the models' inherent differences. Specifically, the comparison was between the Lawrence Livermore National Laboratory SD model (LLM) and the Oregon State University (OSU) two-level GCM. The seasonal variation of the zonal averages of selected climatic variables that are common to both models were compared under conditions in which the insolation and sea-surface temperature were prescribed to follow their observed climatological annual cycles.

Clearly these two climate models will have a different level of accuracy in their simulation of the observed climate. The analysis of such model variability is an important part of model calibration and documentation, but in the reported comparison the aim was to display the models' inherent differences by considering their relative sensitivity or response to a prescribed climatic perturbation. The seasonal cycle of solar forcing was selected as the response against which to examine the relative sensitivities of the two models under standardized conditions.

Description of the models

The OSU GCM is a two-level global model with horizontal resolution of 4° latitude and 5° longitude. At the model's interior σ levels (located at the mid-points of two layers of equal mass between the surface and the model top at 200 hPa), the velocity, temperature, geopotential and water vapour mixing ratio are carried as prognostic variables. Temperature, pressure and surface moisture are, in addition, predicted at the ground. Cloudiness is predicted as a result of either convective or large scale condensation, and a snow cover is possible as part of the simulated surface hydrology. Normally, the model has both diurnal and seasonal variations of solar radiation with sea-surface temperature and sea-ice prescribed from climatology.

A schematic summary of the different vertical structure of the two models is given in Figure 5.7. The LLM two-dimensional SD model has a latitude resolution of 10° and nine isobaric levels in the vertical, five of which are at

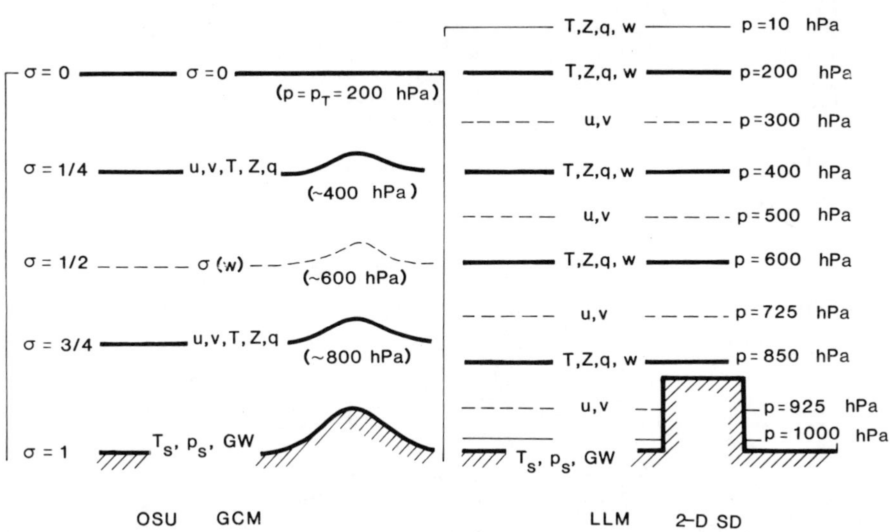

Figure 5.7 The comparative schematic vertical structure of the OSU two-level global GCM (left) and the LLM zonally averaged SD model (right). Here the σ levels of the OSU and the p levels of the LLM are shown (except the levels at 20, 50 and 100 hPa for the SD model), where the temperature (T), geopotential (z), water vapour mixing ratio (q) and the horizontal (u, v) and vertical velocities ($\dot{\sigma}$, w) are predicted. Both models also predict the temperature (T_s), pressure (p_s) and ground wetness (GW) at the surface, although in slightly different ways (reproduced by permission of the American Meteorological Society from Potter and Gates (1984), *Monthly Weather Review*, **112**, 909–917).

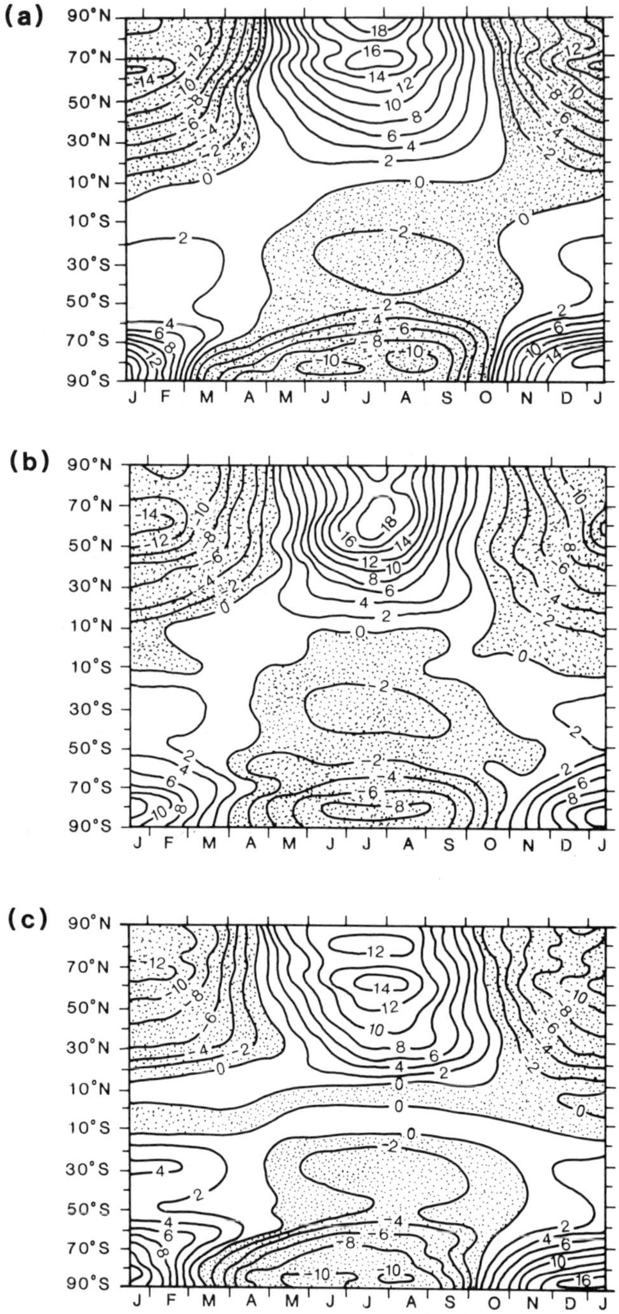

Figure 5.8 Zonal mean seasonal anomaly (difference from annual mean) of surface air temperature (K) (negative values are shaded) from: (a) observations; (b) LLM; (c) OSU (reproduced by permission of the American Meteorological Society from Potter and Gates (1984), *Monthly Weather Review*, **112**, 909–917).

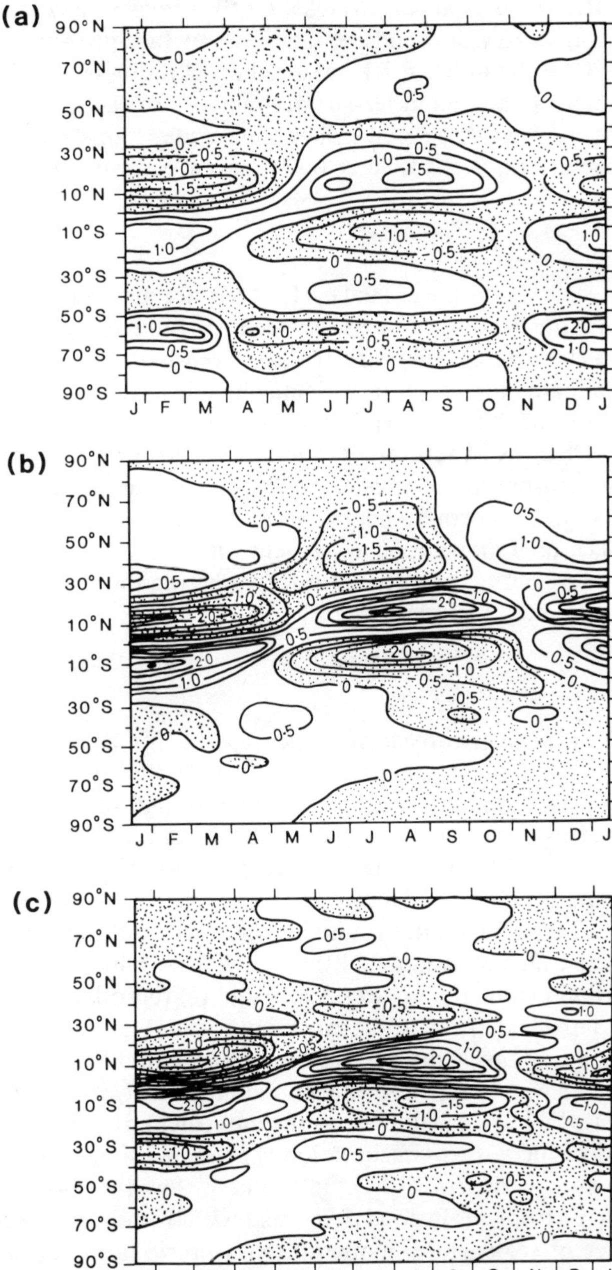

Figure 5.9 Zonal mean seasonal anomaly (difference from annual mean) of precipitation rate (contoured in intervals of 0.5 mm d^{-1}) (negative values are shaded) from: (a) observations; (b) LLM; (c) OSU (reproduced by permission of the American Meteorological Society from Potter and Gates (1984), *Monthly Weather Review*, **112**, 909–917).

or above 200 hPa. The dependent variables are the velocity, temperature, geopotential and mixing ratio (for each layer), and the surface pressure, temperature and humidity. Like the OSU, the LLM predicts cloudiness, precipitation and snow depth. Solar radiation and sea-surface temperature are prescribed. Unlike the OSU GCM, however, regional or synoptic-scale changes are not resolved and the horizontal and vertical eddy transports are parameterized.

Temperature

In Figure 5.8, which shows the annual cycle of surface air temperature departure from the annual mean, both models are seen to simulate the boundary layer temperature in close agreement with the observed values. Such a response is likely to be dominated in the southern hemisphere by the prescribed sea-surface temperatures, and in the northern hemisphere by the seasonal cycle of solar radiation. The OSU GCM reproduces the observed seasonal amplitude of the surface air temperature in the southern hemisphere summer only slightly less successfully than in the northern hemisphere winter. In contrast, the LLM two-dimensional SD model shows good agreement with observations in the northern hemisphere winter, but is less successful in predicting the maximum deviation from the mean in high southern latitudes.

Precipitation

The movement of the intertropical convergence zone (ITCZ) has a marked effect on the observed relative annual cycle of precipitation given in Figure 5.9. The deviations from the annual mean in the middle to high southern latitudes are nearly as great as those in the tropics, yet the total rainfall in the higher latitudes is considerably less. This feature is likely to be due to the uncertainties in the observational data which show summer and winter maxima of 4–5 mm d^{-1} in the southern hemisphere that are not found in some other data sets. Excluding this feature, the relative annual variation of the precipitation which is simulated by the LLM and OSU and shown in Figures 5.9(b) and (c) reproduces the amplitudes which are associated with the seasonal excursion of the ITCZ reasonably well. The LLM produces an extensive dry area in the northern middle and high latitudes which is associated with excessive tropospheric subsidence. The OSU simulates a similar drought which is associated with the movement of the subtropical high, while poleward an increase in precipitation is predicted similar to that observed (which is not shown in the LLM). It is likely that the models' simulated precipitation in the tropics is in good agreement with that observed because of the strong influence of sea-surface temperature on precipitation at low latitudes.

Other features

Both models predict cloud cover although neither reproduces one of the more prominent features in observed cloudiness: the summer maximum cloudiness in high latitudes, particularly in the northern hemisphere, which is attributed to summer stratus. The variation of cloud cover in the LLM two-dimensional SD model in high latitudes is roughly 6 months out of phase with observations, while the OSU GCM performance is not much better. For tropical cloudiness, both models reproduce the seasonal cycle quite well; the simulated subtropical summer cloudiness minimum in the northern hemisphere is extended into middle or even high latitudes. In the case of the LLM, this may be attributed to an excessively intense downward branch of the Hadley circulation resulting from the use of an annual rather than a monthly average eddy momentum transport. The OSU cloud cover, however, occurs where there is precipitation, and so indicates a less severe summer cloudiness minimum than that indicated by the LLM. These predictions of zonally averaged cloudiness could be assessed by comparing top-of-the-atmosphere radiation fluxes with satellite-retrieved values as described in the context of the EBM discussed in Appendix B.

The overall results are that both the two-dimensional SD model and the GCM produce similar seasonal phases and amplitudes of zonally averaged cloudiness, temperature and precipitation in response to the observed variations of incoming solar radiation and sea-surface temperature. Such a similarity of response occurs in spite of the many differences in the models' structure and formulation.

5.6 OTHER TYPES OF TWO-DIMENSIONAL MODELS

5.6.1 An upgraded energy balance model

Naturally it is possible to select two domains from the climate system other than the 'height in the atmosphere' and 'latitude zone' which characterize two-dimensional SD models. For example, the ocean-plus-land 'box' models described in Section 3.6 and Appendix B, could be thought of as two-dimensional models. Indeed a number of modelling groups have 'added' a second dimension to existing one-dimensional models. For example, it is possible to set up one-dimensional RC models for a number of latitude zones as characterized by an EBM. Such a combination would constitute a two-dimensional model.

Another variation on the EBM theme was developed by William Sellers in the 1970s. He worked with the thermodynamical energy equation for the Earth–atmosphere system applied separately to the land- and water-covered portions of each 10° latitude belt of an Earth with a single large continent extending from pole to pole. In other words, this two-dimensional model is a cleverly partitioned EBM. The model was formulated in terms of an idealized continent–ocean system shown in Figure 5.10. Except for latitudes between 40° and 70° S the fraction of each 10° latitude belt occupied by land is equal to the present-day value. Furthermore, the land masses have been offset in relation to one another in order to give a meridional transport across each 10° latitude circle from water to land, water to water, land to land and land to water similar to that currently observed. The energy equation in each area is a function of the surface e.g. the equation for the vertical temperature profile $T(p)$ (p is pressure)

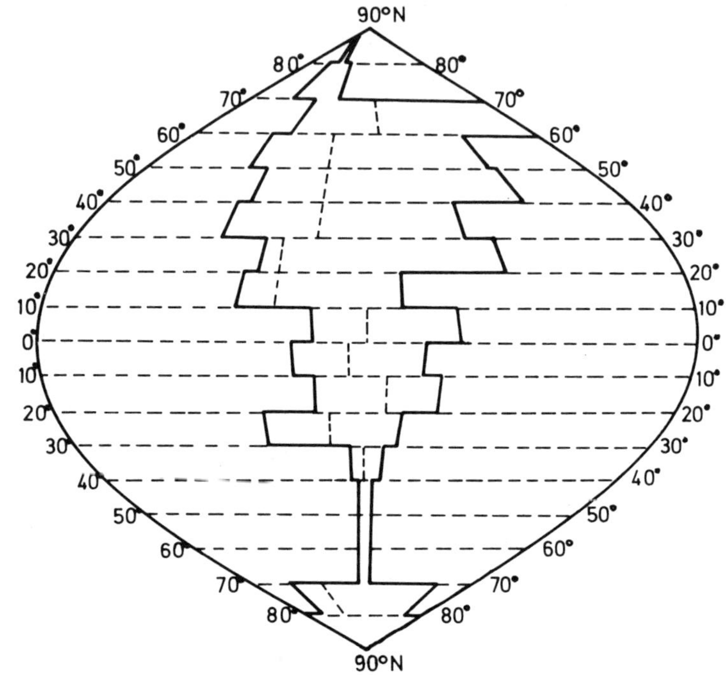

Figure 5.10 Distribution of land and sea within an upgraded EBM. Land masses are offset from each other to simulate the real world case. The land bridge between 60° S and 70° S exists for computational reasons (reproduced by permission of the American Meteorological Society from Sellers (1973), *Journal of Meteorology*, **12**, 241–254).

is written in terms of the surface temperature T_s and the vertical temperature gradient $\partial T/\partial p$:

$$T(p) = T_s - (p_s - p)\ \partial T/\partial p \tag{5.14}$$

where $\partial T/\partial p$ is specified as $0.12\ \mathrm{K\ hPa^{-1}}$.

This two-dimensional model also differs from the basic EBM of Sellers described in Chapter 3 in that the eddy diffusivities are used to parameterize the poleward transport of heat by ocean currents, atmospheric standing waves and transient eddies. The values of the atmospheric eddy coefficients for sensible heat K_H and water vapour K_v (assumed to be equal) are proportional to the first power of the north–south temperature gradient:

$$K_H = K_v = 0.25\,|\Delta T| \times 10^6 \tag{5.15}$$

where ΔT is the temperature gradient between successive 10° latitude belts. In addition an eddy diffusivity K_w for heat transfer by ocean currents was specified as

$$K_w = 1.7 \times 10^6 \times (1 - A_L)A_I \tag{5.16}$$

where A_I is the fractional area of the oceans covered by ice. The factor A_L, being the fraction of a given latitude belt occupied by land, allows for the effect of the continents in channelling the north–south ocean currents. The average value of K_w is about $5 \times 10^4 \, m^2 \, s^{-1}$, which agrees well with values found in the vicinity of the Gulf Stream, which is about one order of magnitude larger than values considered typical for the more quiescent parts of the oceans. This discrepancy was intentional and was an attempt to account partially for the neglected heat transport by vertical circulations (cf. Section 3.6).

The albedo–temperature feedback is still an important feature of this model, but now cloudiness is treated explicitly by computing surface net radiation for the separate cases of clear and cloudy skies. This model is very much more comprehensive than an EBM, but is still highly tuned as compared with the general circulation climate models which will be described in Chapter 6. Despite, or perhaps partly because of, this tuning, this two-dimensional model is able to reproduce many of the fundamental features of the climate system (Figure 5.11).

Despite some discrepancies between modelled and observed values in Figure 5.11, especially in high latitudes, there is a clear indication that this model is able to reproduce seasonal signals in various climatic variables fairly successfully. Although this result might have been anticipated, since much of the climatic response is controlled by thermal inertia, it is still worthy of note because the seasonal range in most, if not all, variables is considerably greater than many anticipated climatic perturbations.

5.6.2 A severely truncated spectral general circulation climate model

An alternative method of achieving a similar position on the climate-modelling pyramid (Figure 2.1), rather than develop or embellish a fundamentally simple zonal model, is to restrict a more complex model. Figure 5.12 illustrates the relative performance (measured qualitatively as a percentage of the dynamical soundness of a GCM) against the computational effort expended. As the difference between the computational domains of spectral and cartesian grid GCMs will not be explained until Chapter 6, suffice it to say here that severe truncation of a spectral model is roughly equivalent to coarsening the spatial grid of a cartesian grid model. In many senses, however, the spectral truncation can be more severe than the equivalent grid coarsening, since when the resolution is decreased in a cartesian grid model, baroclinically unstable waves are more and more poorly resolved, reducing the dynamical soundness of the model simulation. On the other hand, spectral truncation can be very great before the fundamental dynamics are badly misrepresented.

An example of a severely truncated spectral GCM would be one in which there are two layers in the vertical and the zonal spectral representation is of wavenumbers 3, 6 and 9 only (see Section 6.2.2 for explanation). The surface–atmosphere interactions and the surface state are computed in grid-point

(a)

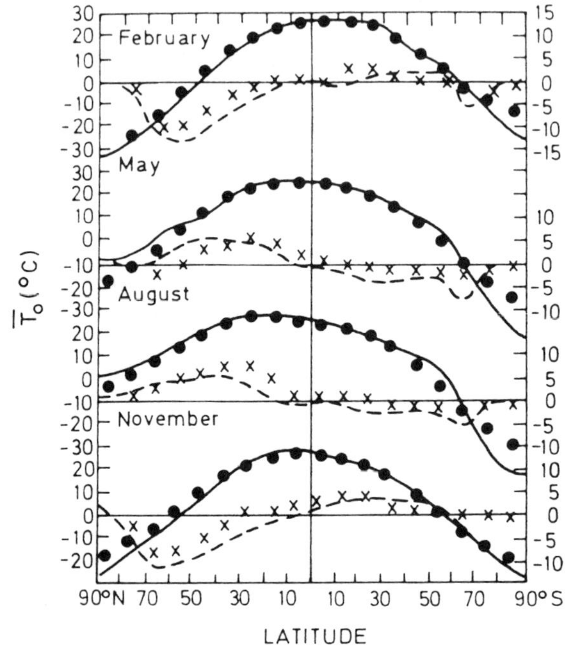
February
May
August
November
$\overline{T}_o$ (°C)
$\overline{T}_{OL} - \overline{T}_{OW}$ (°C)
90°N 70 50 30 10 10 30 50 70 90°S
LATITUDE

(b)

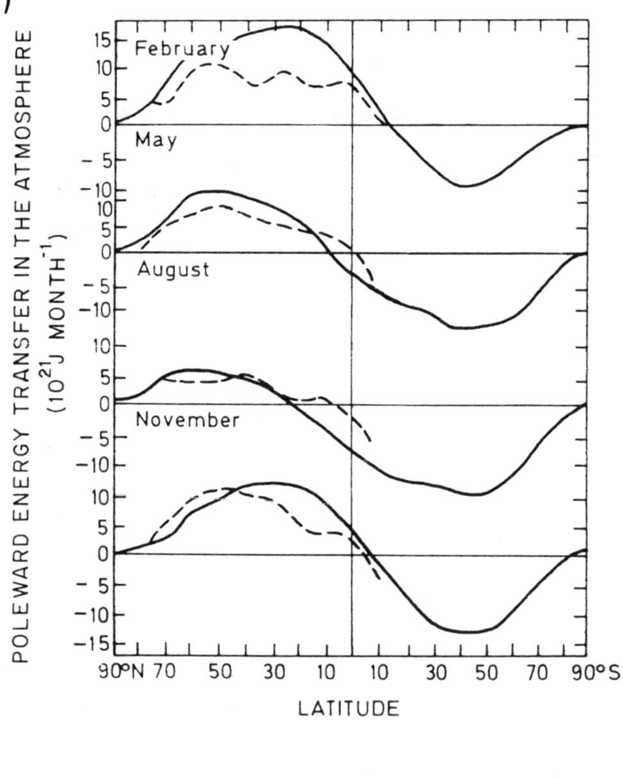
February
May
August
November
POLEWARD ENERGY TRANSFER IN THE ATMOSPHERE
(10^{21}J MONTH^{-1})
90°N 70 50 30 10 10 30 50 70 90°S
LATITUDE

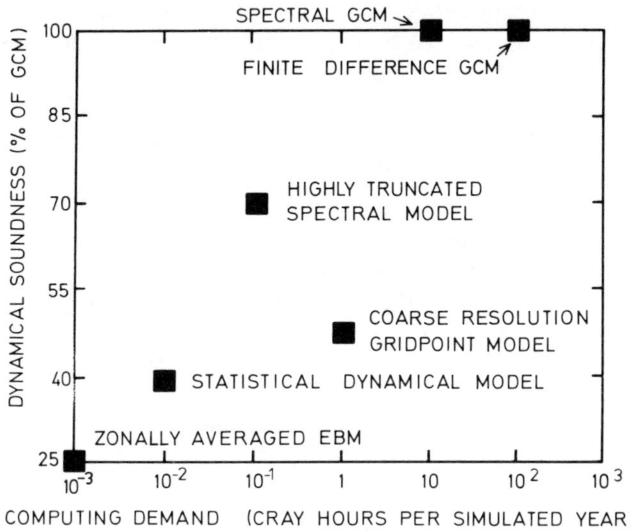

Figure 5.12 Variation of the 'dynamical soundness' of atmospheric models versus their computational requirements. The scale of dynamical soundness is subjective (reproduced by permission of the American Meteorological Society from Semtner (1984), *Journal of Climate and Applied Meteorology*, **23**, 353–374).

space and radiative calculations can be as complex as deemed necessary. This type of highly truncated GCM can be 'better' than a two-dimensional SD model since the large scale eddy fluxes of momentum, heat and moisture are calculated explicitly rather than being parameterized. Despite this explicit computation, the computational effort is approximately two orders of magnitude smaller than that for a fully resolved GCM.

The loss of spectral resolution means that many features are either poorly predicted or must be improved by better sub-gridscale parameterization. Naturally, a very detailed effort here would be inappropriate since the results would be unlikely to better the replacement of lost spectral resolution. Thus this type of model might be considered more appropriate for ocean–atmosphere coupling than, for example, for soil moisture prediction.

It is possible to argue that severely truncated models are similar to two-dimensional models in that they approximate the complete climate system by

---

Caption to Figure 5.11 (*opposite*)

Figure 5.11 (a) Latitudinal variation of the observed mean sea-level temperature, T (solid lines) and of mean sea-level temperature difference ($T_L - T_w$) between land and water (dashed lines) for February, May, August and November. Computed values are indicated by solid dots and crosses (reproduced by permission of the American Meteorological Society from Sellers (1973), *Journal of Meteorology*, **12**, 241–254). (b) Latitudinal and seasonal variation in poleward energy transfer as computed (solid) and observed (dashed) (reproduced with permission from Sellers, 1973).

highly parameterized representation of one or more aspects. On the other hand, as they retain three dimensions they are, perhaps, more appropriately termed two-and-a-half-dimensional models.

Some atmospheric dynamicists have chosen to solve a set of equations similar to equations (5.1)–(5.5) but in three directions (i.e. including longitudinal variations). However, they exclude much of the detail of the treatment of the radiative fluxes which is usual in one-dimensional RC and two-dimensional SD models. Thus, although they are strictly three-dimensional modellers, in choosing to neglect the complexities of radiative processes these modellers have slipped a part-dimension and once again the result might be more honestly termed a two-and-a-half-dimensional model.

As the desire to increase the predictability obtainable with a climate model increases, so in general does the number of dimensions represented. In Chapter 6 the closest point yet attained by climate modellers to the apex of the climate-modelling pyramid, the GCM, is described.

5.7 WHY ARE SOME CLIMATE MODELLERS FLATLANDERS?

In the introduction to this chapter we noted that two-dimensional beings cannot understand three dimensions and must describe what they see in terms of the only dimensions they recognize. Since most climate modellers of our experience are three-dimensional, why is it that some of them choose to represent the climate system in terms of highly parameterized formulations which are essentially a reduction of the dimensionality of the real system?

One answer is, of course, that by simplifying they aim to represent important mechanisms without overburdening their computational resources. In this way many of them can watch their simplified systems develop for longer time periods than would be possible if they were to use GCMs operating on even the best available computers of today. This method can be carried further; for example EBMs have been used to investigate 'Milankovitch' variations in climate. In Section 5.6 we tried to show that the demarcation between two- and three-dimensional models is rather arbitrary. The other reason is that by stripping away 'unnecessary complexities' the workings of any system are more readily viewed and, hence, hopefully more easily understood. Consequently some questions, especially those pertaining to basic mechanisms and to long time periods, are better tackled with simpler (i.e. more highly parameterized) models.

The question of how far two-dimensional models should be used to represent climate scenarios is a difficult one. Recent work has suggested that the *latitudinal* response to an increase in CO_2 and other trace gases is similar whether the perturbation is applied as a single "jump" or as a transient change. On the other hand, preliminary results also suggest that the general spatial character of the transient response (i.e. including longitudinal detail) differs from the steady-state response.

One message that we hope that you have gained from this chapter and that will be underlined in Chapter 6 is that it is not scientifically short-sighted to choose to parameterize some aspects of the climate system and thereby reduce the number of dimensions; it is inevitable. The skill of a good modeller lies in the ability to identify results which are probably features of the real system as opposed to those which are facets of the constrained framework of the model. Thus, climate modellers can choose to dwell happily in Flatland but they must not begin to think like natives.

RECOMMENDED READING

Abbott, E. A. (1963) *Flatland: A Romance of Many Dimensions* (5th edn), Barnes and Noble, New York, 108pp.

Hunt, B. C. (1973) Zonally symmetric global General Circulation Models with and without the hydrologic cycle, *Tellus*, **25**, 337–354.

Oerlemans, J., and Vernekar, A. D. (1981) A model study of the relation between Northern Hemispheric glaciation and precipitation, *Contrib. Atmos. Phys.*, **54**, 352–361.

Ohring, G., and Adler, S. (1978) Some experiments with a zonally averaged climate model, *J. Atmos. Sci.*, **35**, 186–205.

Potter, G. L., and Cess, R. D. (1984) Background tropospheric aerosols: incorporation within a statistical dynamical climate model, *J. Geophys. Res.*, **89**, 9521–9526.

Potter, G. L., and Gates, W. L. (1984) A preliminary intercomparison of the seasonal response of two atmospheric climate models, *Mon. Wea. Rev.*, **112**, 909–917.

Potter, G. L., Ellsaesser, H. W., MacCracken, M. C., and Luther, F. M. (1979) Performance of the Lawrence Livermore Laboratory zonal atmospheric model, in *Proc. GARP JOC Study Conference on Climate Models*, GARP Publication Series No. 22 (Vol. 2), WMO, Geneva, pp. 995–1001.

Potter, G. L., Ellsaesser, H. W., MacCracken, M. C., and Mitchell, C. S. (1981) Climate change and cloud feedback: the possible radiative effects of latitudinal redistribution, *J. Atmos. Sci.*, **38**, 489–493.

Saltzman, B. (1978) A survey of statistical dynamical models of terrestrial climate, *Advances in Geophysics*, **20**, 183–304.

Sellers, W. D. (1973) A new global climate model, *J. Appl. Meteor.*, **12**, 241–254.

Sellers, W. D. (1976) A two dimensional global climate model, *Mon. Wea. Rev.*, **194**, 233–248.

Semtner, A. J. Jr. (1984) Development of efficient, dynamical ocean–atmosphere models for climate studies, *J. Clim. Appl. Meteor.*, **23**, 353–374.

Shutts, G. J. (1983) Parameterization of travelling weather systems in a simple model of large scale atmospheric flow. In B. Saltzman (ed.) *Theory of Climate*, Advances in Geophysics, Vol. 25, Academic Press, New York, pp. 117–172.

Stone, P. H., and Miller, D. A. (1980) Empirical relations between seasonal changes in meridional temperature gradients and meridional fluxes of heat, *J. Atmos. Sci.*, **37**, 1708–1721

Taylor, K. E. (1980) The roles of mean meridional motions and large scale eddies in zonally averaged circulations, *J. Atmos. Sci.*, **37**, 1–19.

Thompson, S. L., and Schneider, S. H. (1979) A seasonal zonal Energy Balance Climate Model with an interactive lower layer, *J. Geophys. Res.*, **84**, 2401–2414.

Chapter 6

General Circulation Climate Models

'Predictability is to prediction as romance is to sex.'
K. Miyakoda (public communication, 1985)

6.1 THE STRUCTURE OF GENERAL CIRCULATION CLIMATE MODELS

General circulation climate models are the 'most complete' models currently available. They model the full three dimensionality of the atmosphere and sometimes include an ocean mixed layer sub-model and a range of other features.

The history of GCMs which was outlined in Chapter 2 is such that they are still often thought of as being composed of three primary components: (i) dynamics, (ii) physics, (iii) other (Figure 6.1). It is perhaps unfortunately also true that the computational effort expended decreases from (i) to (iii).

A typical flow chart for a GCM is shown in Figure 6.2. The *dynamics* comprises the numerical schemes by which large scale atmospheric transports are accomplished. These dynamics are computed either in rectangular grid space or in spectral space as described in Section 6.2. The equations solved are similar to those first formulated for numerical weather forecast models but the GCM schemes have been derived differently to stress, for example, conservation which is much less important for very short-period integrations. The *physics* is usually computed in grid space even in spectral GCMs partly because it pertains to the surface and to vertical motion and partly to limit isolated phenomena, such as thunderstorms, to specific locations. *Other* components can be, for example, linkages between the atmosphere and an ocean circulation model or a sea-ice prediction model.

6.2 DYNAMICS IN GENERAL CIRCULATION CLIMATE MODELS

In this section the basic formulation of GCMs is considered with particular references to the differences between so-called 'spectral' and 'cartesian grid' models. Any GCM must be formulated with some fundamental considerations for:

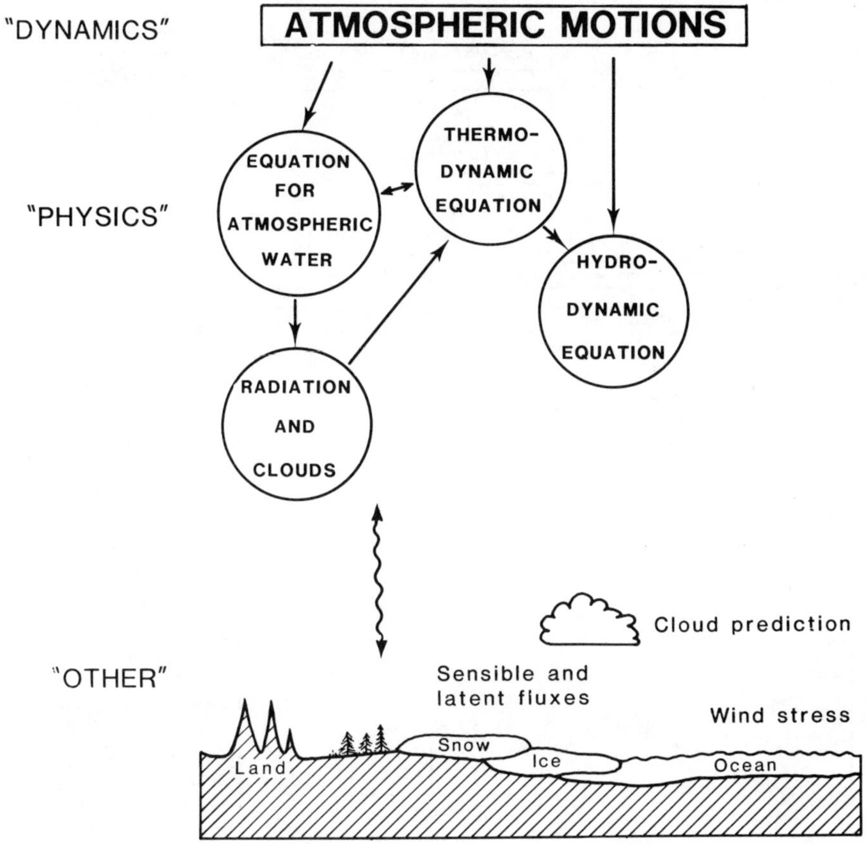

Figure 6.1 The processes incorporated in a GCM. It is generally true that more computational effort is expended on the dynamics and the physics than on the other processes incorporated in GCMs.

1. Conservation of momentum

$$\frac{D\mathbf{v}}{Dt} = -2\Omega\wedge\mathbf{v} - \varrho^{-1}\nabla p + \mathbf{g} + \mathbf{F} \tag{6.1}$$

2. Conservation of mass

$$D\varrho/Dt = -\varrho\nabla.\mathbf{v} + C - E \tag{6.2}$$

3. Conservation of energy

$$\frac{DI}{Dt} = -p\frac{d\varrho^{-1}}{dt} + Q \tag{6.3}$$

4. Ideal gas law

$$p = \varrho RT \tag{6.4}$$

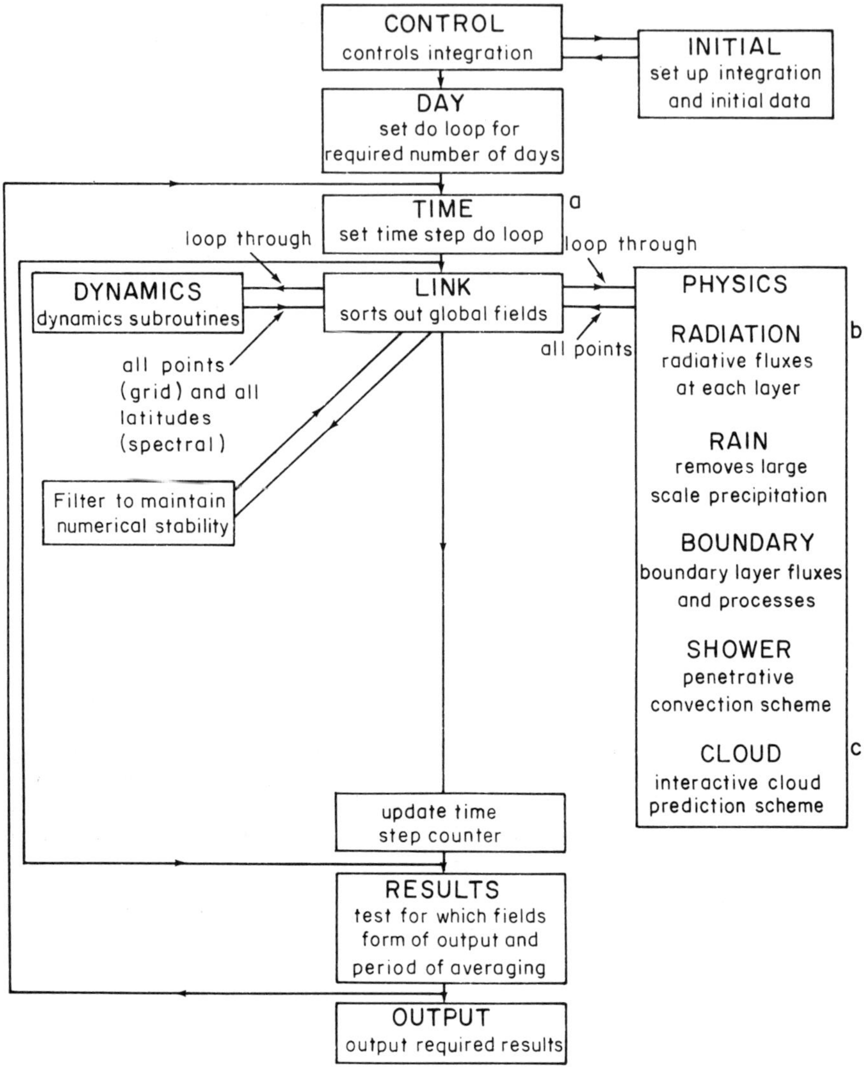

a Generally the time step is about 30 minutes

b RADIATION is not always called at every time step
as dynamical changes are faster than radiative changes

c CLOUDS are often prescribed rather than calculated

Figure 6.2 Generalized flow diagram of a general circulation climate model.

where $\mathbf{v}$ = velocity relative to the rotating Earth,
t = time,

$\frac{D}{Dt}$ = total time derivative $[= \frac{\partial}{\partial t} + \mathbf{v} . \nabla]$,

Ω = angular velocity vector of the Earth,
ϱ = atmospheric density,
$\mathbf{g}$ = apparent gravitational acceleration,
p = atmospheric pressure,
$\mathbf{F}$ = force per unit mass,
C = rate of creation of atmospheric constituents,
E = rate of destruction of atmospheric constituents,
I = internal energy per unit mass $[= c_v T]$,
Q = heating rate per unit mass,
R = gas constant,
T = temperature,
c_v = specific heat of air at constant volume.

A GCM should also conserve enstrophy (the root mean square of the vorticity). Failure to conserve enstrophy means that the energy of motion is transferred unrealistically to smaller and smaller space scales. Very early model structures did not conserve enstrophy and, consequently, became unstable after short integration times.

6.2.1 Cartesian (or rectangular) grid general circulation climate models

Modelling the atmosphere by dividing it into a series of 'boxes' is probably the easiest method to visualize. The atmosphere is reduced to a matrix of numbers, usually evenly spaced in latitude and longitude. Depending on the complexity of the model, values of a number of variables may be stored at each point. In Figure 6.3(a) the structure of a cartesian grid GCM is illustrated. Although the grid boxes are usually regularly spaced horizontally, the vertical spacing of grid boxes is governed by the importance of a particular region of the column. In general there are more layers at the bottom of the column and near the tropopause. In a cartesian grid GCM the structure is as might be expected: variables for a particular location are moved into the computer memory and all computations undertaken. Different grid structures involve the retaining of different variables at different points of the grid. Besides physical considerations there are numerical influences on the grid spacing and timestep of a GCM. The timestep must be short enough that the maximum speed of propagation of information does not permit any transfer from one grid point to another within one timestep. The timestep Δt is therefore governed by the restriction

$$\Delta t \leqslant \Delta x / c \qquad (6.5)$$

(a) CARTESIAN GRID GCM

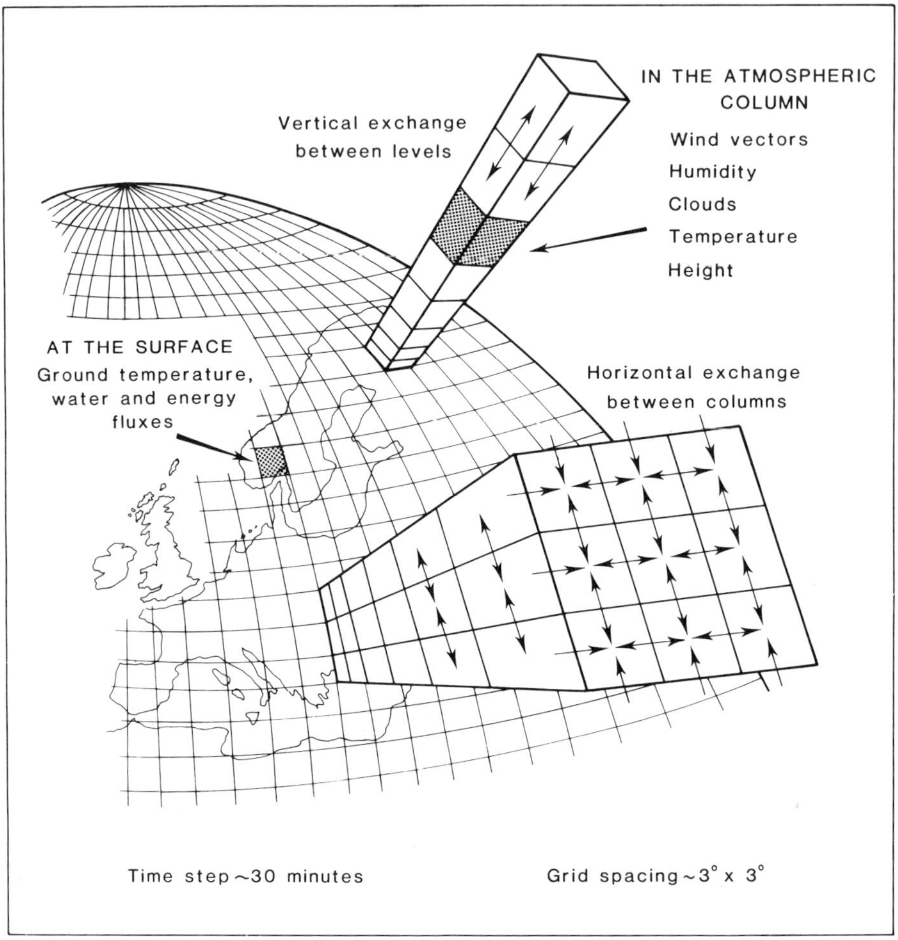

Figure 6.3(a) *caption opposite*

where Δx is the grid spacing and c is the fastest propagation velocity, which in atmospheric GCMs is the speed of gravity waves. In the case of a cartesian grid the grid spacing becomes small at the poles leading to the need for very short timesteps of the order of a few seconds. The incorporation of filtering procedures overcomes the numerical instability caused by not reducing the timestep to match this requirement.

Cartesian grid models are considered more satisfactory than spectral GCMs by many people. There are a number of reasons including the advantage that transfer between two computational systems is not required and the fact that cartesian models do not produce the rather disconcerting predictions of, for example, negative masses of chemical species generated by spectral models. These

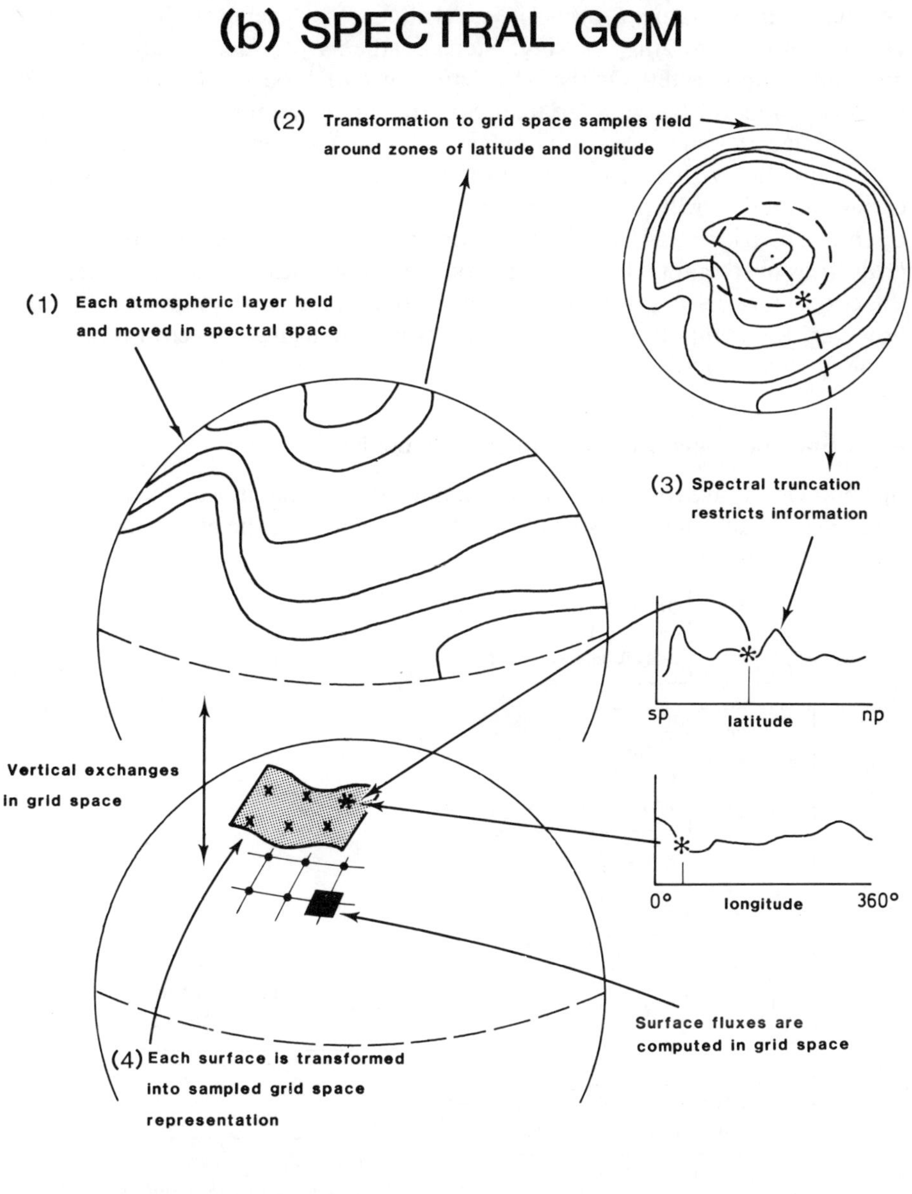

Figure 6.3(b)

Figure 6.3 The construction of: (a) a cartesian (or rectangular) grid GCM and; (b) a spectral GCM. In a cartesian grid GCM horizontal and vertical exchanges are handled in a straightforward manner between adjacent columns and layers. In a spectral GCM vertical exchanges are computed in grid-point space (i.e. in the same way as in cartesian grid GCMs), while horizontal flow is computed in spectral space. The method of transfer between spectral and grid-point space can be followed with reference to the text and reading around Figure 6.3(b) from point 1 to 4.

non-physical phenomena occur in spectral GCMs whenever the considered variable is not smoothly varying. Steep gradients produce negative masses, humidities, etc., locally. On the other hand, spectral models give very successful results for smoothly varying variables. Spectral models require less computational effort than cartesian models of a similar resolution, as is illustrated in Figure 5.12.

It seems likely that as computational power increases and costs decrease, cartesian models will be favoured over spectral models. The next generation of cartesian grid models will be based on locally correct, regional, rectangular grids, thus overcoming the gradient-produced anomalies of spectral models and the coordinate selection difficulties of current cartesian models. At present, however, the computational advantage gained by using a spectral model has resulted in the rapid development of this model type.

6.2.2 Spectral general circulation climate models

Spectral GCMs are formulated in a fundamentally different way from cartesian grid GCMs. Although the surface is retained as a grid, the atmospheric fields

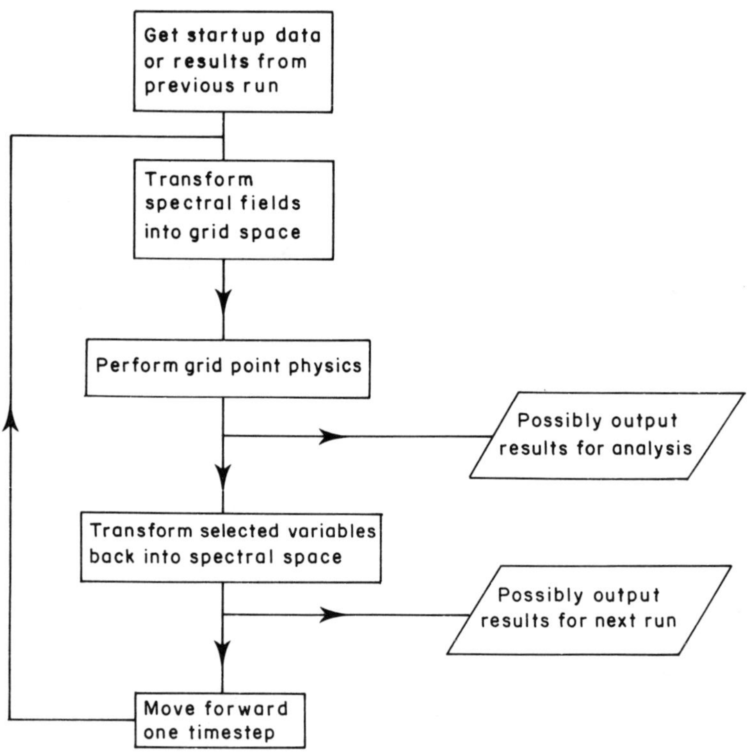

Figure 6.4 The flow through the computational scheme of a spectral GCM. In this case there is only one transformation loop although models with two transformation loops have also been developed.

are held and manipulated in the form of waves (Figure 6.3(b)). The advantage of this is that fewer numbers are required to represent a global field and computation times are hence reduced. Spectral models are, however, not formulated in all directions using waves: a rectangular grid is used for vertical transfers, and radiative transfer and surface processes are modelled in this grid space. The flow of a spectral GCM is illustrated in Figure 6.4. The data fields are transformed to and from grid space at every timestep via fast Fourier transforms and Gaussian quadrature (a form of numerical integration). The timestepping is performed with the waveform representation and grid-point physics are incorporated after the transformation into grid space.

Representing the atmosphere with waves

Fourier's theorem states that any periodic function can be represented as a summation of sine and cosine waves. The variation of any quantity around a latitude zone is necessarily periodic and can therefore be represented as a summation of a number of waves. These waves are the Fourier transform of the original function. In the same way that the logarithm of a number contains the same information as the number so the Fourier transform of a function contains the same information as the original grid representation. Analagously to logarithms, Fourier transformation allows mathematical operations to be performed more easily than if the original form were worked upon.

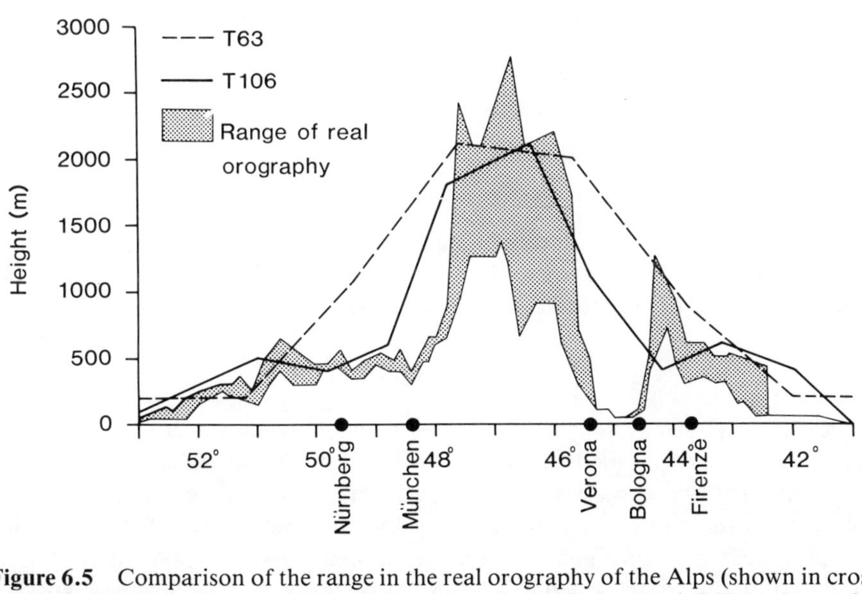

Figure 6.5 Comparison of the range in the real orography of the Alps (shown in cross-section) with the representation used in two versions of a spectral GCM truncated at wavenumbers 63 and 106 respectively (reproduced with permission from Simmons and Bengtsson, 1987).

The same principle of increased computational time being required for increased resolution applies to spectral representation as to cartesian grid models. Resolution in a spectral GCM is governed by the wavenumber of truncation. Typically, for a climate-modelling application around 15–30 waves are used to represent each variable in each latitude zone at each vertical level. If a model has 20 zonal waves then it is said to be 'truncated at wavenumber 20'. An example of the effect of truncation is the smoothing of the real orography of the Alps for two versions of the European Centre for Medium Range Weather Forecasting (ECMWF) model. This model is also spectral in nature, but as it is used for 10–20 day weather forecasts rather than climatic simulations it is truncated at much higher wavenumbers. Figure 6.5 compares the orography for model versions truncated at wavenumbers 63 and 106.

When considering the processes in grid space an array of points, termed the Gaussian grid points, is defined, the number of these points being governed by the particular truncation level of the model. Overspecification of these grid points results in excessive computation times, whereas underspecification results in aliasing (the transferring of energy from high frequencies to low frequencies through poor sampling) of the high frequencies.

Structure of a spectral-type model

We have discussed how a spectral GCM handles information both in wave form and in grid point form. In this section the manner in which information is transferred between these two spaces, being the principal feature of such models, will be described. The full procedure for a single timestep of a spectral model is outlined below.

An arbitrary variable X which has values over the surface of a sphere (e.g. Figure 6.3(b)) can be expanded as

$$X = \sum_{m=-M}^{M} \sum_{n=|m|}^{|m|+j} X_n^m \, Y_n^m \tag{6.6}$$

with X_n^m, the spectral coefficients being complex (i.e. containing an imaginary part premultiplied by $i = \sqrt{-1}$ as well as a real part). The spherical harmonic Y_n^m is a function of longitude, λ, and latitude, ϕ, such that

$$Y_n^m = P_n^m \, (\sin \phi) \, \exp(im\lambda) \tag{6.7}$$

where P_n^m is a Legendre polynomial of degree n and zonal wavenumber m. This is the case at point 1 on Figure 6.3(b). The nature of this functional representation of the atmospheric variables is governed by the spherical coefficients. In the next stage of the model cycle (point 2) the spectral coefficients of the dynamics variables (vorticity, divergence, wind components, water vapour mixing ratio and pressure) are transformed to a latitude–longitude grid (called the Gaussian grid). First a Legendre transform is evaluated for each spectral variable at each of the Gaussian latitudes, ϕ_j. These latitudes are related to the resolution of the model such that they are the roots of the associated Legendre polynomial of order zero. For a variable X,

$$X_n^m \rightarrow X_{(m)}(\phi_j;\, t) = \sum_{n=|m|}^{|m|+j} X_n^m\, P_n^m \tag{6.8}$$

The resulting Fourier harmonics, $X_{(m)}$, at each of the Gaussian latitudes ϕ_j and at time t are transformed via a fast Fourier transform (FFT) algorithm to longitudes $\lambda_l = 2\pi l/L,\ 1<l<L$

$$X_{(m)} \rightarrow Z(\lambda_l,\, \phi_j;\, t) = \sum_{m=-M}^{M} X_{(m)} \exp(im\lambda_l) \tag{6.9}$$

By these two steps we have achieved a spatial distribution of values at a series of grid-points. Physical processes, such as radiative transfer, convection and the ground energy budget can now be calculated in this rectangular grid (point 3 in Figure 6.3(b)). Subsequent to such calculations, variables are transformed back into wave space via fast Fourier transform, and then the inverse Legendre transformations are built up, one latitude at a time, using Gaussian quadrature (Figure 6.4) such that

$$Z_n^m = \sum_{j=1}^{k} w_{j,k}(\phi_j) Z_{(m)}(\phi_j) P_n^m(\sin \phi_j) \tag{6.10}$$

With the model reassembled in spectral form and appropriate expressions for the time rate of change of the variables incorporated then some form of timestep procedure is applied to find the values of the spectral fields at the advanced time point. This returns us to point 1 in Figure 6.3(b). Timestepping can be accomplished in several different ways, but all rely on approximating the time differential in a finite difference form. This gradient is applied for the appropriate time period and a value for each variable is derived at the next time point.

Truncation

The type of truncation of a GCM is often quoted in model descriptions. The simplest explanation is that the truncation number represents the number of waves which are resolved around a latitude zone. This is the governing factor in determining the resolution of a spectral GCM. More completely it is a description of the relationship between the largest Fourier wavenumber, the highest degree of the associated Legendre polynomial and the highest degree of the Legendre polynomial of order zero. These are termed M, K and N respectively. With reference to Figure 6.6, the truncation types are defined as

Triangular: $M = N = K$ (Figure 6.6(b))

Rhomboidal: $K = N + M$ (Figure 6.6(c))

Trapezoidal: $N = K > M$ (Figure 6.6(d))

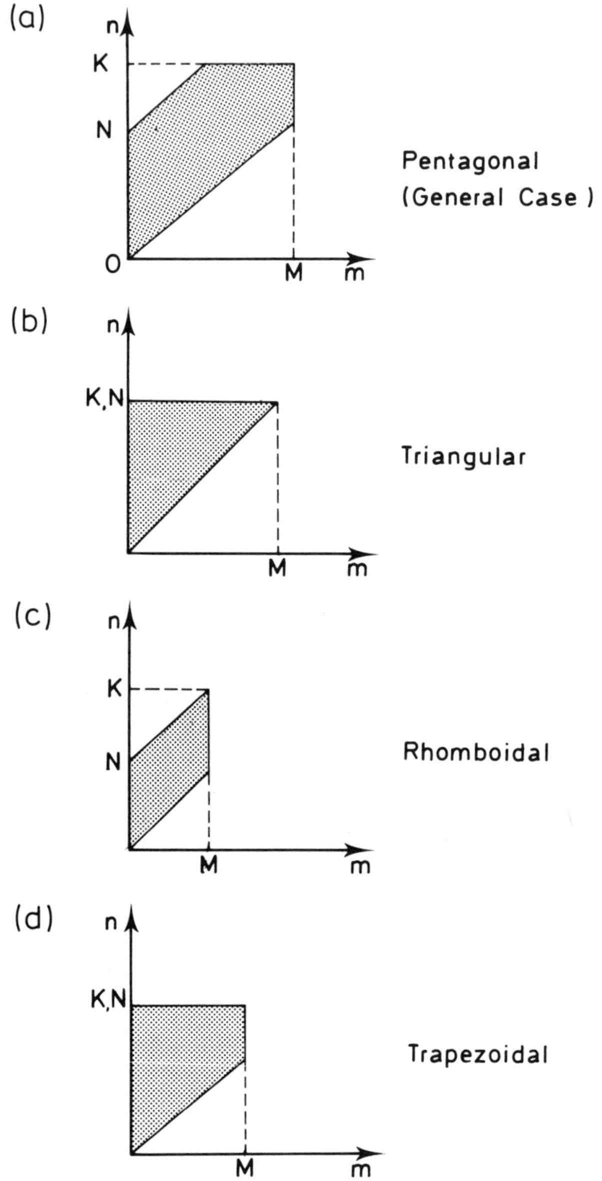

Figure 6.6 The relationship of spectral variables for different truncation types used in spectral GCMs. The most common truncation types are triangular and rhomboidal.

These are all subsets of the pentagonal case illustrated in Figure 6.6(a). These truncation types carry with them requirements for the resolution of the Gaussian grid of the model. For example, for triangular truncation

$$N_{\text{long}} \geqslant 3M + 1; \qquad N_{\text{lat}} \geqslant (3M + 1)/2 \tag{6.11}$$

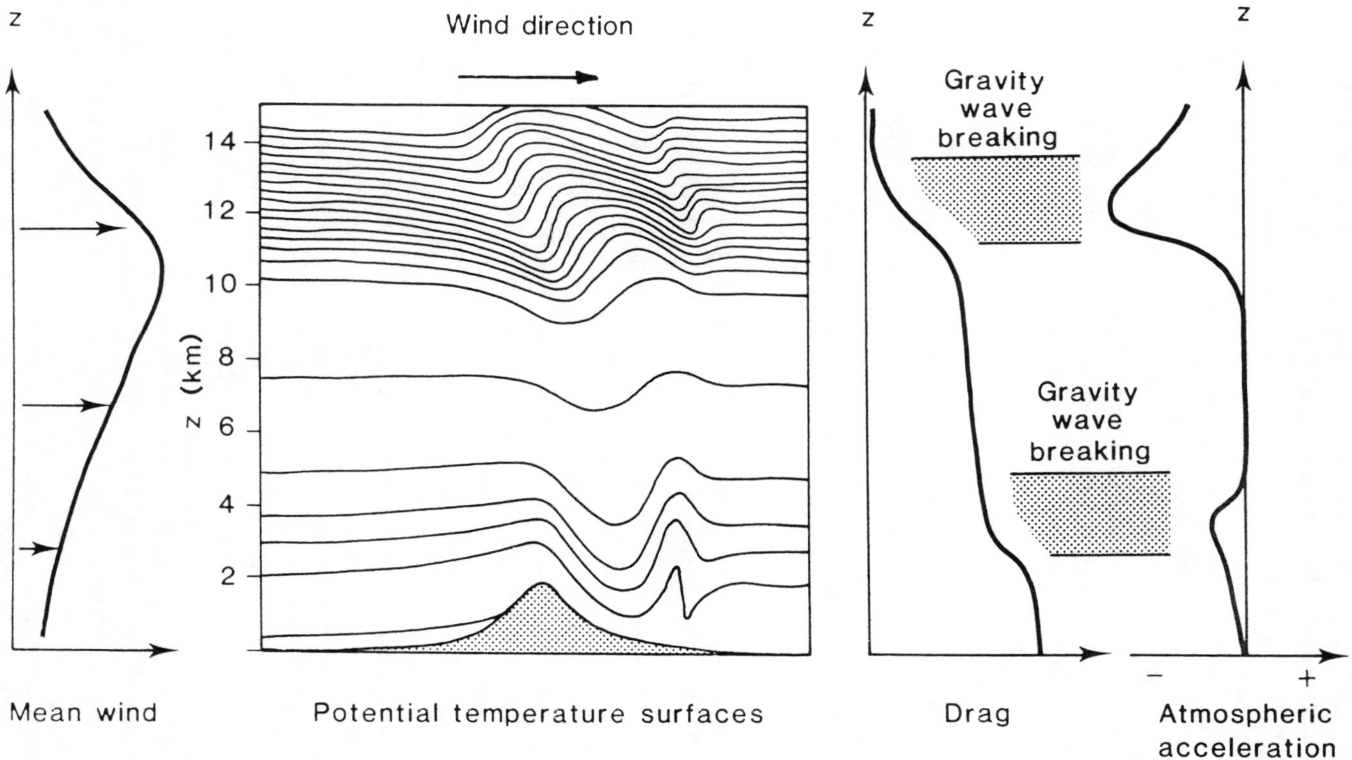

Figure 6.7 Schematic representation of the cause, impact and effect on drag and atmospheric motion of gravity waves induced by an orographic obstacle. Atmospheric gravity waves behave in a manner analogous to water waves in the ocean.

Such a specification is sufficient to remove most of the aliasing in the model.

The two most common truncations used in atmospheric GCMs are the triangular and rhomboidal methods. The choice of truncation is somewhat arbitrary. Although rhomboidal truncation gives higher resolution in high latitudes and has some advantages for vectorization on some supercomputers and can be considered a more 'exact' representation, triangular truncation is generally considered satisfactory and gives a more uniform representation.

Cartesian models, which were the first type to be developed, have been overtaken, somewhat, by spectral models. However, as noted above, future developments in computing may mean that cartesian models become relatively more efficient than the presently preferable spectral models.

Gravity wave drag

Gravity wave drag is the drag of the mountains on the atmosphere which is manifested in the production of gravity waves. This is illustrated schematically in Figure 6.7. These gravity waves can break in a manner analogous to waves on a beach and can transfer momentum from large scale flow to smaller scale flow. These waves have recently been recognized as an important feature which requires inclusion in order to simulate the global and regional scale circulations.

6.3 PHYSICS IN GENERAL CIRCULATION CLIMATE MODELS

The use of σ coordinates combined with the assumption that the atmosphere can be considered to be in hydrostatic equilibrium permits the vertical wind component, w, to be obtained diagnostically from the convergence of horizontal wind components, u and v. In GCMs three-dimensional, time-dependent equations govern the rate of change of the six basic model variables: surface pressure, two horizontal wind components, temperature, moisture and geopotential height. The six basic equations solved to derive these variables are, as described in Sections 5.2 and 6.2, the hydrostatic equation, two equations for horizontal motion (equation (6.1)), and the thermodynamic (equation (6.3)), water vapour and mass continuity (equation (6.2)) equations.

For the purpose of computation of the physics, all GCMs can be visualized as consisting of an array of columns, which extend into both the atmosphere and the surface, distributed over the globe on a regular latitude–longitude grid (Figure 6.3(a)). Separate subroutines are used to solve the dynamical equations; and other subroutines for other physical processes. The horizontal grid used in model integrations varies but a typical resolution is 3–5° latitude by 4–7° longitude. Modelling groups often use variable resolution models so that, for example, a coarser version may be developed for the coupled ocean–atmosphere experiments.

The atmospheric column above each surface point is divided into a prespecified number of layers. Finer vertical resolution occurs near the tropopause and the surface. The near bottom layers are influenced by boundary layer processes.

High resolution near the surface is required to determine temperature and humidity gradients and atmospheric stability, needed for modelling surface fluxes.

Owing to the divergence of solutions at alternate timesteps, a time filter must often be employed to prevent computational instability. Additionally, a multipoint filter is applied in the east–west direction in the polarmost rows. The need for smoothing close to the poles is peculiar to cartesian grid schemes and is not required in spectral GCMs. At each timestep the physics subroutines are used. A typical timestep length is 10–30 minutes.

Generally only a handful of subroutines is used to model physical processes (Figure 6.2). Typical subroutines might be:

1. The radiation scheme;
2. The boundary layer scheme;
3. The surface parameterization scheme;
4. The convection scheme;
5. The large scale precipitation scheme.

All of these schemes, with the possible exception of the radiation scheme, are used at each location, each timestep.

6.3.1 Radiative transfer

The radiation scheme of a GCM is likely to incorporate both daily and annual solar cycles. The radiation schemes are, in general, less complex than those incorporated in one-dimensional RC models (see Section 4.3) but are broadly similar. For computational economy the radiation scheme is often called only every few hours rather than every timestep. The mean cosine of the solar zenith angle over the interval between calls to the radiation scheme is calculated at each point and used to obtain an accurate estimate of incoming solar radiation over the period. In many GCMs cloud amounts are prescribed rather than being calculated. If this is the case, specification is often of zonally averaged climatological cloud amount which is likely to have been collated from a number of different sources. The cloud amount is updated by interpolation between the available data (usually monthly means).

In order to minimize the number of possible pathways of radiation while at the same time adequately representing the major cloud radiation processes, four cloud types can be selected. Layer clouds may occur as low, middle or high cloud, and convective clouds may also be allowed.

Layer cloud is usually assumed to occupy the whole depth of the model layer in which it occurs and convective cloud will be modelled as a vertical cloud column. No reflection of radiation or thermodynamic interaction by the sides of convective clouds is allowed with the surrounding clouds or air, and they are allowed to extend through more than one model layer. Shortwave cloud properties are generally prescribed (e.g. Table 4.1) and usually no account is taken of the dependence of cloud properties on cloud thickness or solar zenith angle. Mixing ratios for ozone are derived from climatological zonally averaged values and are typically prescribed and updated monthly.

In a GCM, radiative fluxes are calculated explicitly from the temperature, humidity and prescribed (or predicted) cloud and ozone distributions. Carbon dioxide is usually assumed to be uniformly mixed and to have a global concentration of ~330 ppmv. The shortwave and longwave fluxes are, of course, treated separately. Example radiation schemes are described briefly below.

Shortwave radiation

Solar radiation is absorbed and scattered by atmospheric gases, clouds and the Earth's surface. As a result of the computer time limitations on GCMs, shortwave absorption by minor constituents and reflection by clouds and the surface may have to be treated in a single spectral band. In such a case, absorptivity curves for water vapour, carbon dioxide and ozone are taken from observational data. Pressure scaling (the effect of pressure on the amount of absorption) is applied only to water vapour amounts. Suppose the incident solar beam is split into five components in passing through the atmosphere (one direct and four diffuse beams, one from each cloud type). Any radiation entering a cloud is attenuated according to prescribed cloud properties such as those in Table 4.1 and the remainder emerges from the cloud base as diffuse radiation. Absorption by minor atmospheric constituents is calculated for each component. The gaseous pathlength of the diffuse beam is determined by multiplying the direct pathlength by the diffusivity factor, 1.66 (see Section 4.3.1). The pathlength for direct radiation is simply a function of the solar zenith angle. Rayleigh scattering is incorporated implicitly by reducing the solar radiation by a suitable amount (~3 per cent). The solar constant used in models takes a range of values, with 1370 W m^{-2} being typical. A simple treatment of enhanced surface absorption due to multiple reflection between clouds and the ground is likely to be included. The effective surface albedo can thereby be expressed as

$$\alpha_{\mathrm{eff}} = \alpha(1 - CR)/(1 - \alpha CR) \qquad (6.12)$$

where α is the true surface albedo, C the mean cloud cover and R the mean reflectivity of clouds as seen from the surface. The value of α_{eff} is calculated by weighting each cloud reflectivity by the amount of solar flux reaching the ground from that cloud.

Longwave terrestrial radiation

The treatment of longwave radiation absorption is likely to be more complex than that for shortwave. For example in one such scheme seven spectral divisions are used. The first interval includes the 6.3 μm vibration–rotation water vapour band and the far infrared rotation band of water vapour; a further two intervals include the overlap between water vapour and the 15.0 μm carbon dioxide absorption bands. The fourth and fifth bands include the effect of the 9.6 μm ozone band and the water vapour continuum. The emissivities and transmissivities for each gas in each interval at some temperature, e.g. 263 K, are sometimes stored in the computer code as 'look-up tables'. The surface and clouds are usually assumed to act as black bodies for longwave calculations, although in this example high cloud amounts are often halved to allow for the low optical thickness of cirrus cloud. After calculation of radiative fluxes a possible way of checking the calculation is to compare the top-of-the-atmosphere fluxes with those derived from satellites, for example data from the Earth Radiation Budget Experiment (ERBE), could be used. A simple zonal example of this is given in model 3 described in Appendix B.

6.3.2 Boundary layer

The atmospheric boundary layer is the region in which surface friction has a large effect on the flow; typically the lowest kilometre or so. In particular, the near-surface wind is backed from the wind direction in the free atmosphere creating the Ekman spiral (Figure 6.8(a)). This layer can also suffer large fluctuations in temperature and humidity (Figure 6.8(b)) and its depth changes over the diurnal cycle. These features cannot be fully represented in most GCMs primarily because the vertical resolution is inadequate and the parameterization schemes are unable to produce adequate approximations to the processes involved. There is a wide range of boundary layer schemes used in GCMs. Typically the interaction between the surface and the lowest model layer is similar in all these schemes, but they differ in their treatment of the turbulent exchanges between model layers in the boundary layer. An example of one such scheme (used in the UK Meteorological Office 11 layer AGCM) is outlined below.

When a call is made to the boundary layer subroutine the surface temperature, snow depth and soil moisture content are first adjusted to accommodate changes in surface temperature due to radiative increments. The potential temperature difference $\Delta\Theta(z)$ between the surface and the lowest model level, at height z, and the difference $\Delta q(z)$ between the saturated specific humidity at the surface temperature and pressure and the specific humidity of the lowest layer are calculated.

A bulk Richardson number, R_i, is then calculated. This is dimensionless: a measure of the stability of the lowest model layers and is a function of $\Delta\Theta(z)$, $\Delta q(z)$, the temperature of the lowest layer, T, and the wind speed $V(z)$ of the lowest layer, such that

$$R_i = \frac{gz[\Delta\Theta(z) + 0.61T\,\Delta q(z)]}{TV^2(z)} \tag{6.13}$$

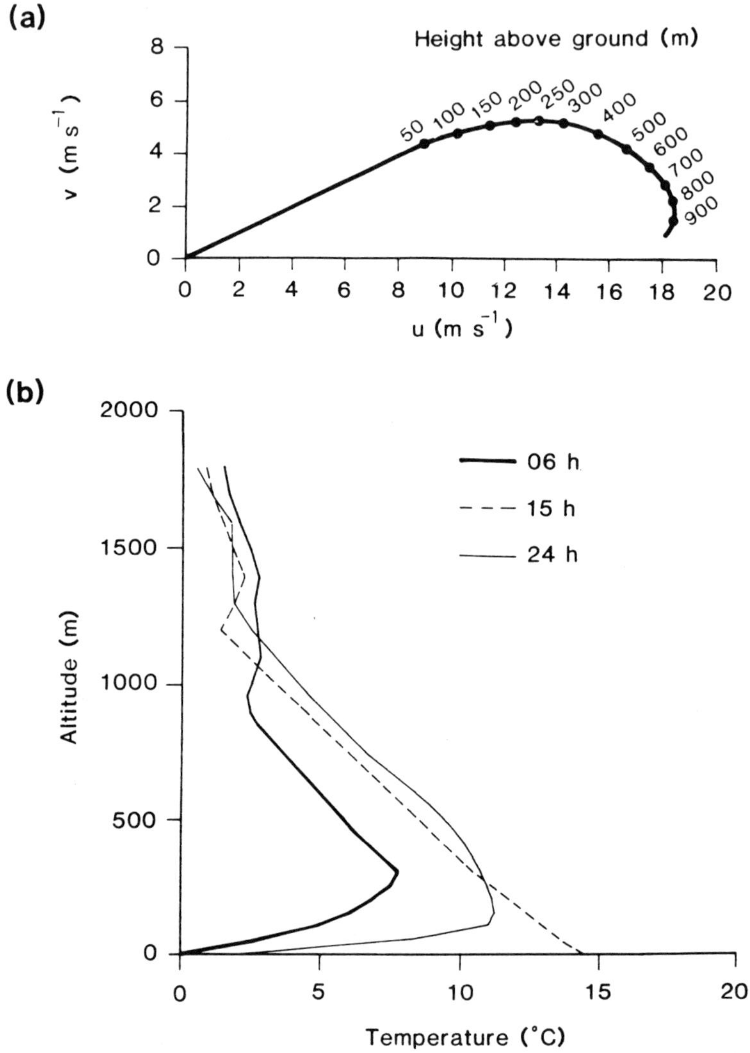

Figure 6.8 (a) The Ekman wind spiral plotted as a function of the two horizontal wind components and; (b) a winter diurnal temperature variation in the planetary boundary layer. Typically neither of these phenomena can be fully represented in GCMs because the vertical resolution in the models near the surface is too coarse. For example, an 11-layer vertical spacing such as that depicted in Figure 4.2(a) has only two layers in the lowest kilometre so that temperature and wind profiles are represented by two points.

The calculated value of the bulk Richardson number is used in conjunction with a specified surface roughness length to determine the bulk transfer coefficients for momentum, sensible heat and moisture. Values of surface roughness are specified. For example, they could be taken to be 0.1 m over land and 10^{-4} m over sea. The surface flux, F_x, of any variable x is given by

$$F_x = -C_x V(z) \Delta x(z) \tag{6.14}$$

where C_x is the bulk transfer coefficient and $\Delta x(z)$ the difference between the value of x at height z and at the surface. The surface fluxes are calculated and the temperature and moisture content of the surface and lowest atmospheric layer are updated accordingly.

For soil moisture content values, S_{mc}, less than some critical value, S_{crit}, evaporation is reduced as follows:

$$\text{Evaporation} = \text{Potential evaporation } (S_{mc}/S_{crit}) \tag{6.15}$$

Snowmelt occurs when the radiative increment would increase the surface temperature of snow-covered land above 273 K and the surface temperature and snow depth are updated.

The turbulent exchanges between atmospheric layers in the boundary layer are modelled using the concept of eddy diffusivity. It is assumed that the turbulent flux, F'_x, between adjacent model levels is proportional to the vertical gradient of that quantity and is given by

$$F'_x = -K_x (\mathrm{d}x/\mathrm{d}z) \tag{6.16}$$

where K_x is the diffusion coefficient for property x. The diffusion coefficient above can be expressed

$$K_x = l^2 (\mathrm{d}V/\mathrm{d}z) \tag{6.17}$$

where V is the horizontal wind speed and l is the mixing length.

6.3.3 Surface parameterization

The surface parameterization is often very simple, with a surface type being prescribed for each grid area and the surface represented by only one layer. Typically at least five possible surface types are included: sea, sea-ice, land-ice, snow-free land and snow-covered land. For each surface type, albedo is prescribed and for types other than snow-free and snow-covered land, soil moisture content is fixed or varies according to some relatively simple parameterization. Over land, a maximum moisture capacity, often termed by modellers the 'field capacity', is specified. A widely accepted value is 15.0 cm. The soil moisture content is updated according to precipitation and evaporation totals. All precipitation is assumed to run off if soil moisture content becomes equal to the specified field capacity. This soil moisture 'book-keeping' is known as the 'bucket' or Budyko model (see Section 6.5). The sea-surface temperature and sea-ice extent are often derived from climatologies and these variables are prescribed every few days by linear interpolation between observed seasonal values.

6.3.4 Convection

The convection schemes used in GCMs are fairly similar. The major components of a typical scheme are outlined here. The scheme is based on the concept of a rising thermal of buoyant air, which is modified by entrainment of environmental air as well as by detrainment. It is assumed that each model layer may produce an ensemble of buoyant thermals which extend upwards to different heights depending on the temperature and humidity characteristics of individual thermals. Entrainment rates of environmental air are prescribed for different model levels. If, on reaching the new level, the virtual potential temperature of the thermal ensemble is less than that of the environment, then ascentof a fraction of the ensemble of thermals is terminated, and detrainment may occur. In this way the virtual potential temperature of the thermal ensemble is raised to slightly above that of the environment. The thermals ascend until a level is reached at which they no longer have a buoyancy excess over the environment

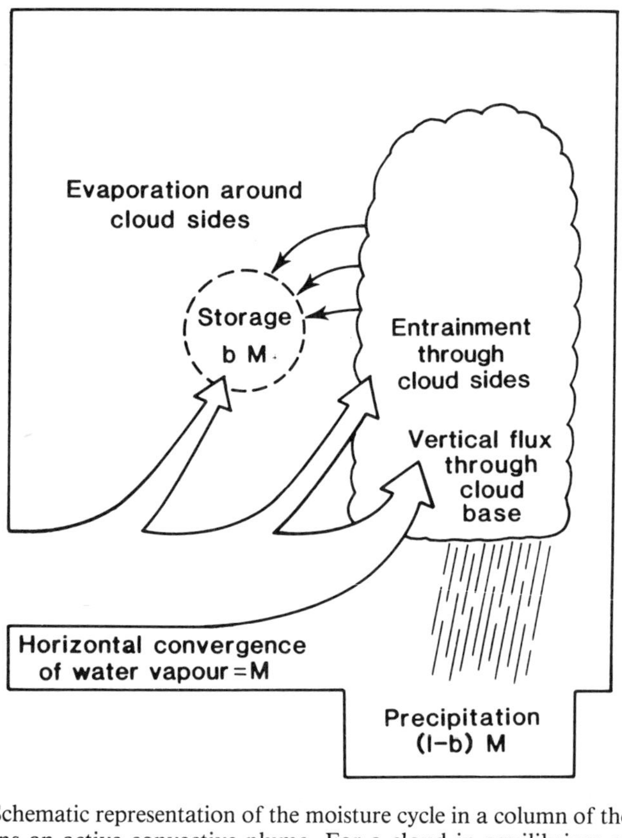

Figure 6.9 Schematic representation of the moisture cycle in a column of the atmosphere which contains an active convective plume. For a cloud in equilibrium a fraction, b, of the mass of water vapour, M, converging into the column is evaporated and the remainder is precipitated (reproduced by permission of the American Meteorological Society from Anthes (1977), *Monthly Weather Review*, **105**, 270–286).

and then ascent ceases. As the thermal rises and cools, any liquid produced by condensation falls as convective precipitation. To allow for the effect of cloud dissipation and unsaturated downdraughts, evaporation of a specified proportion of the condensed water may occur in the layer immediately beneath that which produced it. These processes are summarized in Figure 6.9. The major difficulty is parameterizing the sub-gridscale nature of convection. Often convection would be occurring in the real atmosphere over part of a $5° \times 5°$ area but the average conditions would not satisfy the convection criterion. Similarly, the height of penetration of convection may often be less than the vertical distance between GCM layers. Some parameterization schemes use random number generators to try to simulate sub-gridscale processes.

6.3.5 Large scale rainfall

The large scale rainfall is usually based upon the saturation vapour pressure of each layer. Moisture is permitted to condense out of supersaturated air. If the temperature of the lowest atmospheric layer is less than 273 K the precipitation is classed as snowfall and the snow depth incremented accordingly. If the lowest layer temperature exceeds 273 K, precipitation occurs as rainfall and soil moisture content is incremented.

6.4 INCLUDING 'OTHER' ELEMENTS IN GENERAL CIRCULATION CLIMATE MODELS

In this section we consider some of the features which have been grouped as 'other' processes. Incorporation of these processes into GCMs is becoming more common as modelling abilities and computational power increase. Figure 6.1 is not a precise representation of any one specific GCM but illustrates the typical relative commitment of effort to dynamics, physics and the inclusion of 'other' features.

6.4.1 Cloud prediction

Clouds in GCMs are often specified either zonally or globally. In this section we consider the manner in which cloudiness may be incorporated as a predicted variable. There are basically two types of cloud: large scale supersaturation cloud and convective type clouds. We shall deal with convective clouds initially.

The simplest convective scheme is the adiabatic convective adjustment, as described in Chapter 4. A problem arises with GCMs because the area of a convective cloud and of convective processes is much less than the size of a GCM grid unit. This is the same problem as described in the context of two-dimensional SD models in Section 5.3.

The extent of the differences among currently used cloud prediction schemes is highlighted in the context of a comparison of the performance of three GCMs

Table 6.1 Comparison of the cloud prediction schemes used in three GCMs (after Schlesinger and Mitchell, 1985)

GISS (Goddard Institute for Space Studies)

Clouds are allowed to form in each layer below 100 hPa due to large scale condensation and cumulus convection. A large scale cloud can occur when RH > 100%; cloud cover $C=0$ if $f_s<N_r$ and $C=1$ if $f_s>N_r$, where f_s is the saturated fraction of a gridbox obtained under the assumptions that the absolute humidity is uniform throughout the gridbox and the temperature has a prescribed Gaussian sub-gridscale form, and N_r is a generated random number between 0 and 1. A convective cloud can occur when there is cumulus convection; cloud cover $C=0$ if $\gamma M<N_r$ and $C=1$ if $\gamma M>N_r$, where M is the mass of saturated air rising through the base of a layer and $\gamma=2.5\ \mathrm{m^2\,s\,kg^{-1}}$. Optical depth for large scale clouds $\tau=1/3$ for $T<258$ K and $\tau=0.0133\ (p-100)$ for $T>258$ K, where T is the layer temperature and p is the level pressure in hPa. For convective clouds $\tau=0.08\ \Delta p$, where Δp is the cloud pressure thickness in hPa. Cloud absorptivity and emissivity are derived from phase function and single-scattering albedo based on Mie computations with radii of 10 μm and 25 μm for water and ice clouds respectively.

NCAR (National Center for Atmospheric Research)

Clouds are allowed to form in each layer above the lowest layer and below 250 hPa for $|\phi|<45°$ and below 400 hPa for $|\phi|>45°$ latitude due to large scale condensation and cumulus convection. A large scale cloud occurs when RH > 80% with a cloud cover $C=0.95$. Convective cloud occurs when $\partial\Theta_e/\partial z<0$ and RH > 80% with $C=0.30/N$ in each of the N layers undergoing moist convective adjustment. Clouds in different layers are assumed to be randomly overlapped. Cloud albedo $\alpha=\beta\mu/(\beta+\cos\zeta)$ where ζ is the zenith angle, $\mu=\min(1, \sqrt{[LWC/100]})$ with LWC ($\mathrm{kg\,m^{-2}}$) the column condensate during 1 hour of model time, $\beta=0.6$, 0.3, 0.15 for low ($0.811\leqslant\sigma\leqslant0.991$), middle ($0.5\leqslant\sigma\leqslant0.811$) and high ($\sigma<0.5$ and $p>250$ hPa for $|\phi|<45°$ and $p>400$ hPa for $|\phi|>45°$ latitude) respectively. For clouds occurring simultaneously in two or more layers, the albedo is taken equal to the maximum over all the layers; cloud absorption only by water vapour and ozone. Emissivity $\epsilon=1$ for high cloud and $\epsilon=\mu$ for middle and low clouds.

GFDL (Geophysical Fluid Dynamics Laboratory)

Clouds are allowed to form in each layer. Large scale cloud occurs when RH > 99% with a cloud cover $C=1.0$. Convective cloud occurs when $\partial\Theta_e/\partial z<0$ and RH > 100% with $C=1.0$. For thin clouds (existing in one layer only), the albedo and absorptivity for $\lambda<0.7\ \mu$m ($\lambda>0.7\ \mu$m) are 0.21 (0.19) and 0.0 (0.04) for high cloud (> 10.5 km), and 0.45 (0.35) and 0.0 (0.2) for middle cloud (~ 4.0–10.5 km), 0.65 (0.55) and 0.0 (0.30) for low cloud (0.0–4.0 km). For thick cloud (existing in two or more contiguous layers), the albedo and absorptivity are 0.57 (0.47) and 0.0 (0.3). The emissivity is 0.6 for high cloud and unity for middle and low cloud.

in a numerical simulation of the climatic impact of doubling atmospheric CO_2. The numerical and physical parameterization schemes used in the National Center for Atmospheric Research (NCAR), the Geophysical Fluid Dynamics Laboratory (GFDL) and the Goddard Institute for Space Studies (GISS) GCMs have been compared in Table 6.1.

All three models form clouds by both large scale condensation and convection. Large scale clouds are predicted when the relative humidity exceeds 80, 99 and

100 per cent in the NCAR, GFDL and GISS models respectively, with the amount of coverage being set at 95, 100 and 100 per cent respectively; although the GISS model assigns cloud cover in the grid box for only a certain percentage of the time (the percentage being based on a scheme which compares the fraction of the grid box that is saturated to a random number generated in the interval 0–1). Convective cloud is predicted when there is moist convection, although the GISS model actually generates convective cloud only when another test based upon a random number generator is satisfied. Convective cloud is 100 per cent in the GFDL and GISS models and $30/N$ (per cent) per layer in the NCAR model, where N is the number of layers undergoing convection.

The optical properties of the clouds depend on the cloud height in all three models; in the NCAR model optical properties are also a function of the solar zenith angle and in the GISS model the optical depth is a function of temperature, pressure and pressure thickness (see Table 6.1). The treatment of terrestrial and solar radiation is otherwise very similar in all three models, although the value of the solar constant differs, being 1367, 1370 and 1467 W m^{-2} in the GISS, NCAR and GFDL models respectively. Only the GISS model includes the diurnal as well as the seasonal cycle. The resulting layer cloud amount is produced globally but may be biased by other model characteristics such as the specification of sea-surface temperature or of ocean mixed layer depth, sea-ice extent and soil moisture variability.

It is important to recognize that cloud prediction in GCMs requires sub-gridscale parameterization: cumuliform clouds are significantly smaller than the grid size of GCMs and stratiform condensation is likely to occur over smaller vertical distances than the vertical resolution of the GCM. The radiative properties are also prescribed at present in such a way that, often, the solar and infrared properties are incompatible. This latter problem can be overcome if both short and longwave cloud properties are calculated as a function of a single variable such as cloud liquid (or frozen) water content. The difficulties of sub-gridscale parameterization are much harder to overcome but improvements are only likely to be gained when more detailed observational data become available.

Figure 6.10 shows the latitude versus height distribution of cloud amount as predicted by these three different GCMs. The differences among these predicted distributions are quite considerable, especially in high latitudes. There are also large differences between the vertical distribution of the predicted clouds. These may be more important than simple total cloud amount differences as high clouds tend to warm the system while low and middle level clouds tend to cool.

6.4.2 Modelling the cryosphere

Any complete model of the climate must include prediction of snow cover, sea-ice and possibly permafrost. General circulation models attempt to simulate all aspects of the climate system although the accuracy with which they can do this is limited by several factors.

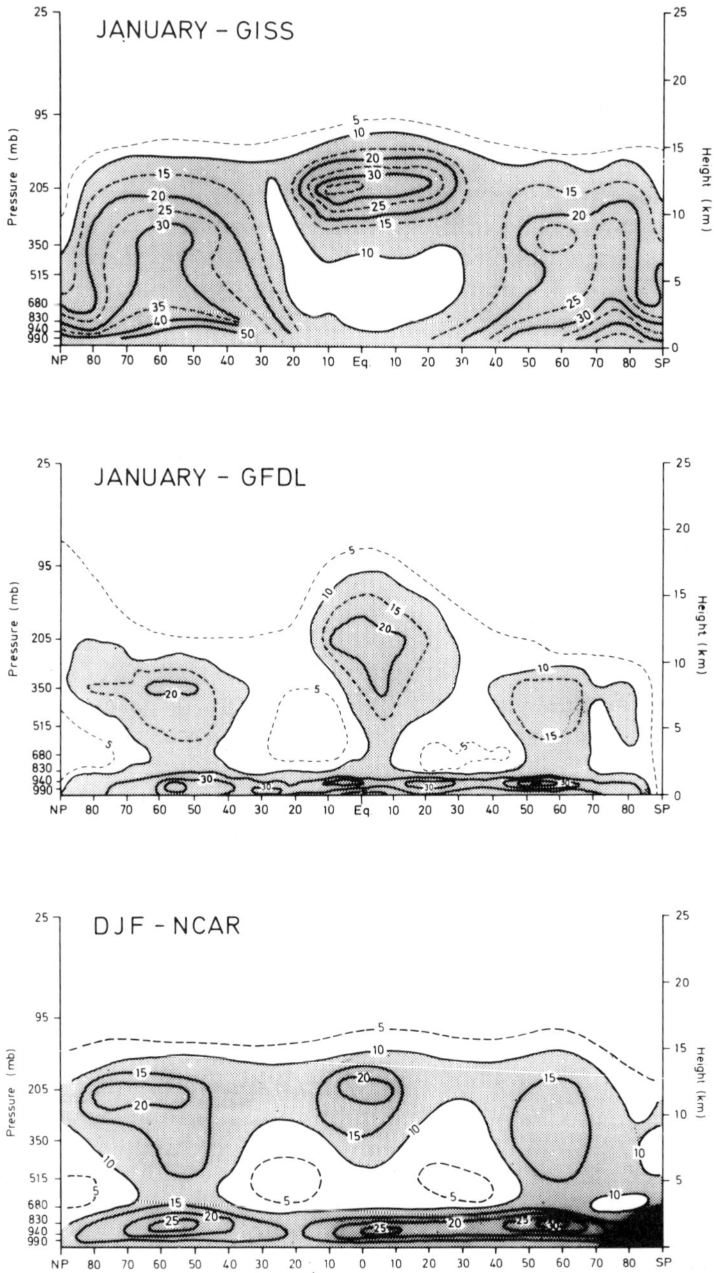

Figure 6.10 (*caption opposite*)

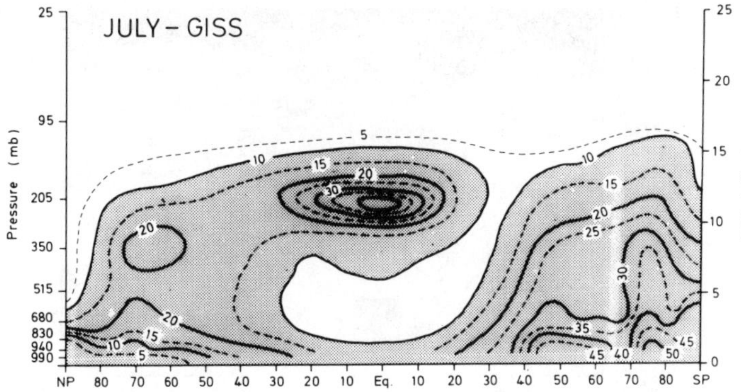

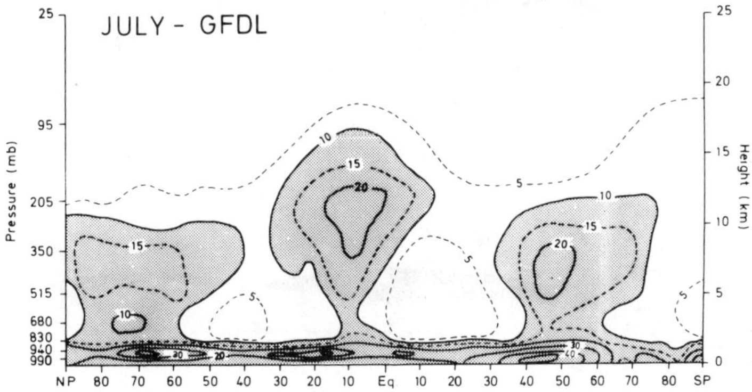

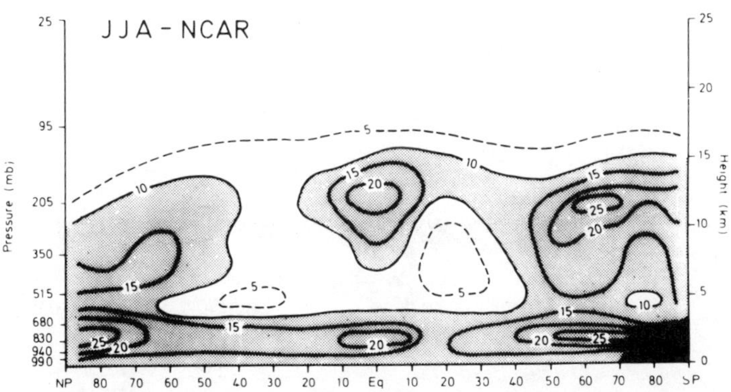

Figure 6.10 Vertical distribution (zonal mean) for cloud predicted from three different general circulation climate models: GISS, GFDL and NCAR. Simulations are shown for January (or December, January, February) and July (or June, July, August) (after Hansen, personal communication, 1983; Manabe and Wetherald, 1982, and Washington and Meehl, 1983 respectively).

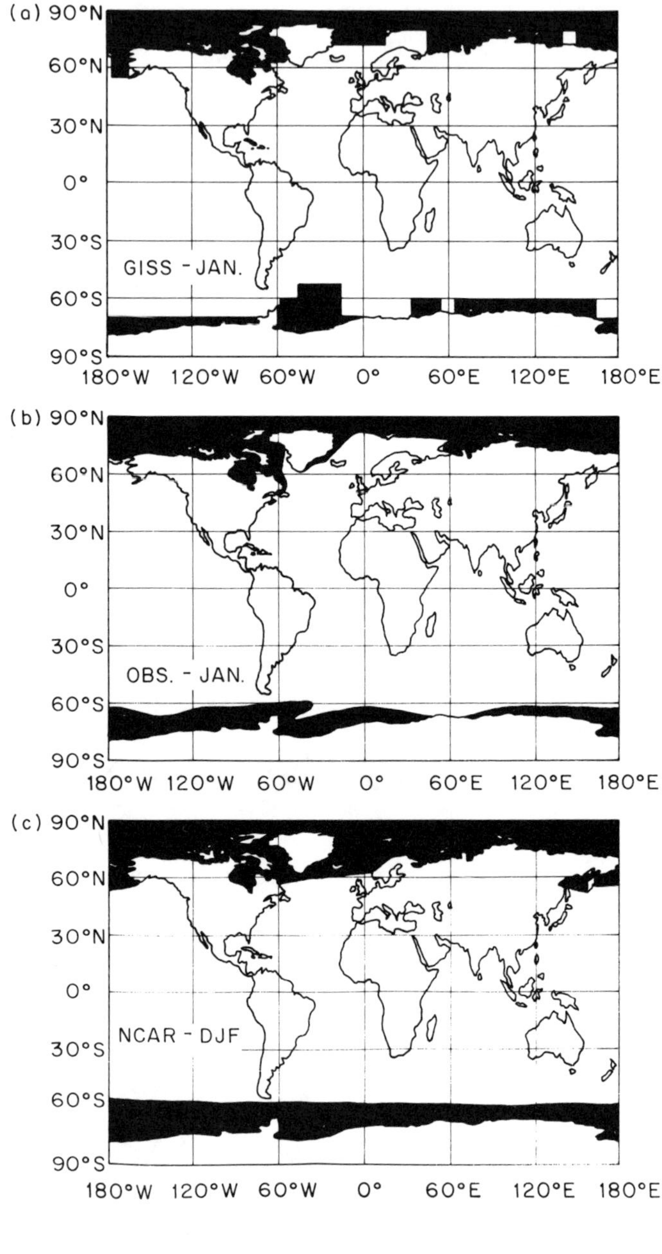

Figure 6.11 (*caption opposite*)

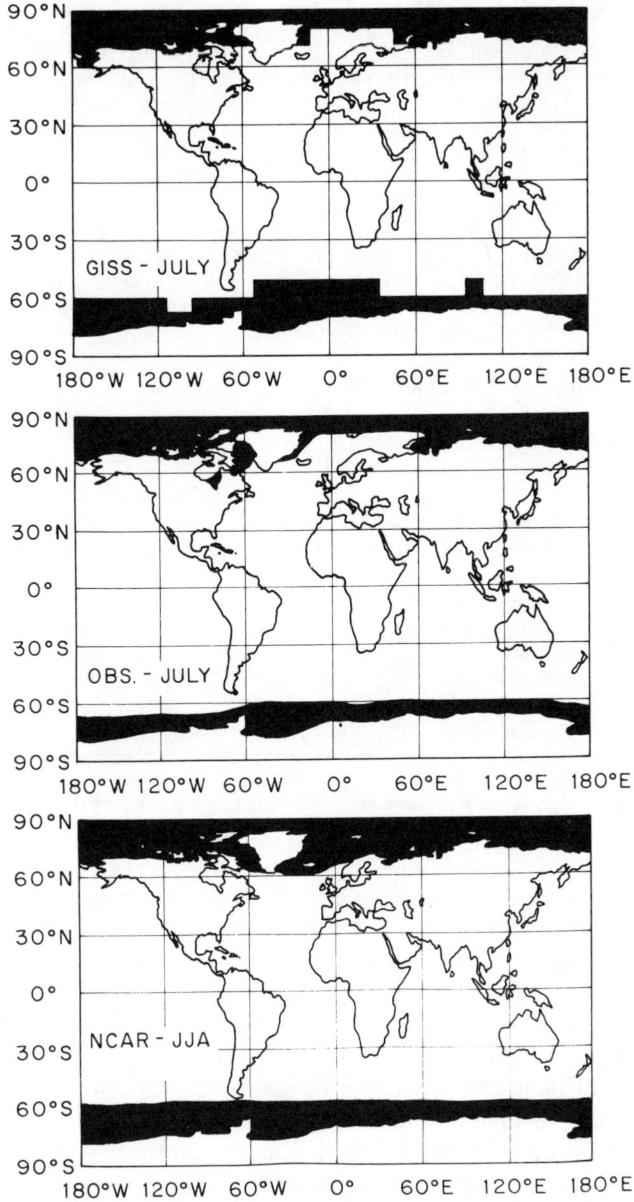

Figure 6.11 Predicted sea-ice extent from two coupled atmosphere–ocean climate models: (a) reported by Barry *et al.* (1984); and (c) after Washington and Meehl (1984). Simulations are shown for January (or December, January, February) and July (or June, July, August). Observations from Alexander and Mobley (1974) are given in (b). The cartesian grid model (a) is seen to produce a more realistic prediction of sea-ice extent despite its coarser resolution. This is because this model includes (prescribed) ocean transports.

Table 6.2 Range of parameterization of ice surface and snow albedos in GCMs

Model	Sea-ice albedo (T = surface temp.; ϕ = latitude)	Snow albedo (T = surface temp.; ϕ = latitude)
GLAS	Always 0.7	Always 0.7
ANMR (Australia)	Always 0.7	Latitudinally varying
GFDL (Version 1)	0.7; $T \leqslant 263$ K 0.35; $T > 263$ K	$\alpha_1 + (0.6 - \alpha_1)d_{sw}^{1/2}$ for $d_{sw} < 1$ cm 0.6 for $d_{sw} \geqslant 1$ cm
GFDL (Version 2)	0.5; $\phi < 55°$ 0.7; $\phi > 66.5°$ 0.45 if surface melting	As above except that 0.6 is replaced by a latitudinally dependent deep snow albedo
UK Met. Office	0.8; $T < 271.2$ K 0.5; $T \geqslant 271.2$ K	$\alpha_1 + 0.38d_{sw}^{1/2}$ $\alpha \leqslant 0.6$ (sea-ice + permanent snow cover) 0.8 ($T < 271.2$ K) 0.5 ($T \geqslant 271.2$ K)
NCAR	0.5 and the addition of snow raises this bare ice value	$\alpha = 0.85$, $\lambda \leqslant 0.7\ \mu$m $\alpha = 0.55$, $\lambda > 0.7\ \mu$m
GISS	0.45 bare ocean ice, higher if snow covered, max. for deep total cover is 0.85	$0.5 + \exp(-\text{age}/5)$ (partly covered) $\alpha_1 + (\alpha_s\text{-}\alpha_1) \times (1 - \exp(-d_{sw}/d_{mask})$ two spectral regions
CCSAS (USSR)	Always 0.6	$0.2 + 0.4d_{sw}$ but $\leqslant 0.6$
UCLA	Bare ice 0.4	Snow-covered ice 0.7

α_1 is the albedo of snow-free land. α_s is the fully snow-covered albedo; d_{sw} is the water equivalent depth of snow (cm); d_{mask} is the vegetation snow-masking depth equivalent thickness of water (cm).

All climate models at present neglect the effect of permafrost entirely. Sea-ice is widely modelled in GCMs, but with only limited success since models are restricted to a simple thermodynamic computation. Snowfall and snow cover are variously predicted in GCMs based on ambient temperature and precipitation predictions. Temperate glaciers and all glacier dynamics are neglected in climate models.

Many of the difficulties associated with successful incorporation of cryospheric elements into GCMs result from the fact that parameterization depends upon successful prediction of other features. This dependence is well illustrated in Figure 6.11 which compares summer and winter predictions of sea-ice extent made using a cartesian grid model (Figure 6.11(a)) and a spectral model (Figure 6.11(c)). Despite the fact that the spectral model has a higher effective spatial resolution (~4° × 7°) than the cartesian model (8° × 10°) and the assertion in Chapter 5 (Figure 5.12) that spectral models generally outperform cartesian models, it can be seen that the spectral model overpredicts sea-ice extent as compared with observations (Figure 6.11(b)). This is because the spectral model does not include ocean transport while the cartesian model has fixed ocean transports. The effects of neglecting oceanic energy transport are most clearly seen in the region of the Gulf Stream.

As the dominant component of the summer-time surface energy balance in cryospheric regions is solar radiation, it is essential that the large scale surface albedo be parameterized correctly. However, the sudden decrease observed in Arctic sea-ice albedo does not occur on the Antarctic pack ice. Conditions in the Antarctic are not favourable for the formation of melt puddles which means that the summer-time decrease of albedo in the Southern Ocean is much less than near the North Pole. It is possible that one of the reasons why the coupled ocean–atmosphere–ice models described above underpredicted circum-Antarctic sea-ice extent in the summer could be the use of albedos significantly lower than those observed in this region, i.e. the use of Arctic albedos in the Antarctic region.

The albedo of sea-ice is predicted variously in GCMs (Table 6.2). For example one parameterization is: $\alpha = 0.5$ if latitude $< 55°$, $\alpha = 0.7$ if latitude $> 66.5°$, with linear interpolation for intermediate locations. If the ice is melting the albedo is reduced to 0.45 and when thickness is less than 0.5 m albedo decreases as a square root function of thickness until it equals the albedo of the underlying surface. The ice thickness is predicted as the balance of freezing and melting at the ice bottom and of melting, sublimation and precipitation at the top. This complex albedo parameterization of the GFDL model contrasts starkly with the specification of a sea-ice albedo of 0.7 in some other models.

In models using specified ice cover, the effect of such differences in albedo will be minimized since anomalies in surface radiation balance will not be able to feed back on the ice amount or thickness. This will not be the case with coupled GCMs.

The prediction of snowfall in GCMs can be generalized as follows. Predicted precipitation will fall as snow if the bottom layer of the atmosphere has a temperature below 0 °C. If the surface is above 0 °C then the snow melts, cooling the surface and adding to ground water or runoff. Snow prediction is therefore dependent on the ability of the model to simulate the hydrological cycle and the distribution of surface and air temperatures. The effect of snowfall depends on incorporation of the effect of vegetation. For example, snow falling on tundra can raise the albedo from 0.2 to 0.8, whereas the same snow cover on a dense coniferous forest may not raise the albedo at all in the long term because of the forest canopy. Also, the modelling of snow surface albedos in GCMs differs (Table 6.2) and thus even similar predictions of snowfall and snow extent could lead to different radiative effects.

Two further aspects of the climate system are of considerable interest to most GCM modelling groups. The land surface and the oceans had been neglected until very recently, but both areas have undergone significant developments in recent years. The ocean, despite covering over two-thirds of the Earth's surface, has until recently been parameterized as a swamp or at best a relatively shallow mixed layer without horizontal advection. The land surface had received still less attention until model simulations of, for example, the effects of increasing atmospheric CO_2 focused attention upon the possibility of water stress and altered growing season length. In Sections 6.5 and 6.6 some of the recent developments in land surface parameterization and in coupling ocean and atmospheric models are described.

6.5 LAND SURFACE PARAMETERIZATION IN GENERAL CIRCULATION CLIMATE MODELS

Surface hydrology

It is widely agreed that surface processes are one of the more important aspects of climate. Over 70 per cent of the energy absorbed into the climate system is absorbed by the surface, and experiments with many climate models have underlined the sensitivity of the climate to surface hydrology. Two difficulties which are currently hampering progress will be discussed. A major problem is the reconciliation of basin hydrological studies with the resolution typical of general circulation climate models: $5° \times 5°$ grid elements. The second, and acute, problem is the dearth of hydrological data with which to initialize and validate global models.

Various GCM studies have identified the importance of the surface hydrology for climatic simulations. For example, Figure 6.12 shows the considerable impact on rainfall of modifying surface evaporation. In this experiment, evaporation from the land-surface was forced to be equal to the potential evaporation (free evaporation) or set equal to zero at all land surface points. As can be seen, for the month of July the resulting precipitation is vastly reduced (shown in Figure 6.12). As well as the considerable reduction, there is also a shift in the position of the remaining maxima of rainfall over the continental areas. This experiment, although extreme, suggests that the modelling of evapotranspiration from the land-surface may be crucial to the accurate modelling of the global hydrological regime.

Despite these findings, the land-surface hydrology has, traditionally, been incorporated into GCMs using exceedingly simple parameterizations; the Budyko or 'bucket' model being a good example. In this parameterization the soil is considered to be a bucket having some maximum depth, usually termed by the modellers 'field capacity' (Figure 6.13(a)). The bucket fills when precipitation exceeds evaporation and when it is full, excess water runs off. It is also important to note that runoff in GCMs cannot play any further role in the hydrological cycle of the model. Although more generalized and more sophisticated bucket models have been developed by various GCM groups (e.g. the two-soil-layer model shown in Figure 6.13(b)), they can be, at best, only approximations to the surface processes since, for example, more than half of the water lost to the atmosphere from the soil flows through plants, rather than simply from the soil.

Global data sets

Currently available computer-readable global data sets are rare. There are a

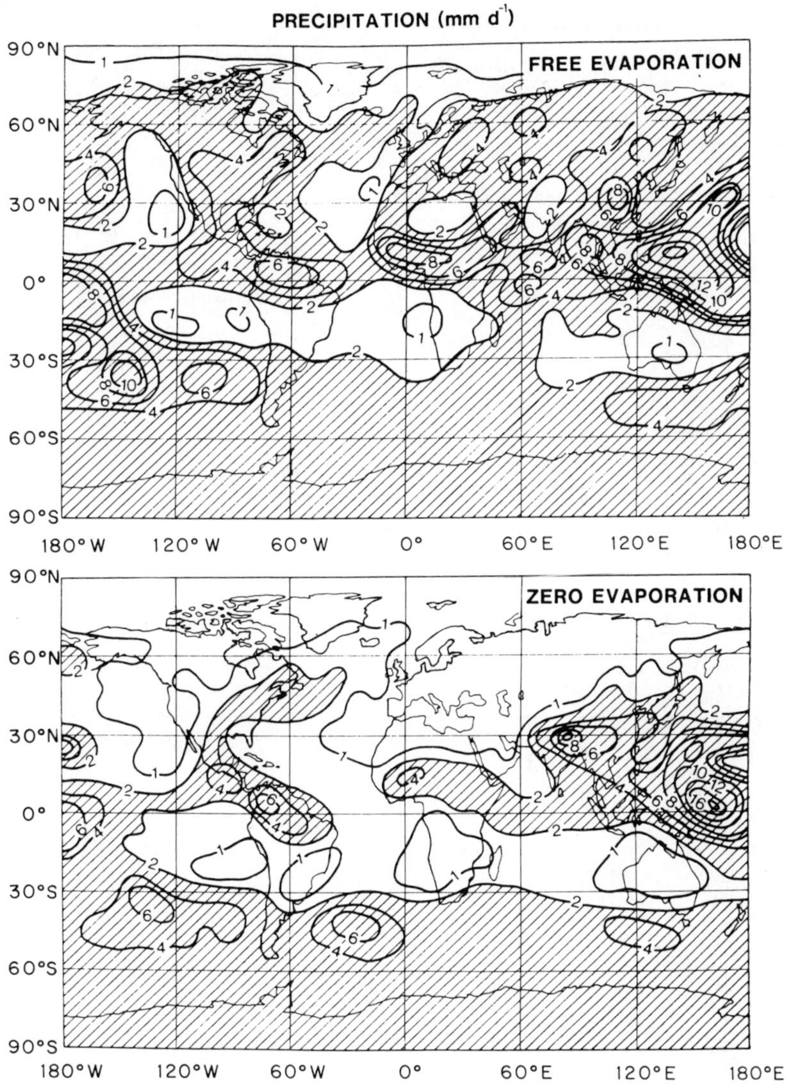

Figure 6.12 The effect on the precipitation of reducing to zero the evaporation from the continental land surface as compared with permitting evaporation at the potential rate as simulated by a general circulation climate model. Results are averaged from the last 30 days of two 60 day 'perpetual July' simulations (after Shukla and Mintz, 1982).

few land-use/vegetation archives. Most of those currently available to the global climate-modelling community have been compiled primarily from atlases, although very recently meteorological satellites (in particular the advanced very high resolution radiometer (AVHRR) sensor of the NOAA polar orbiting satellites) have provided global-scale information about some land-surface

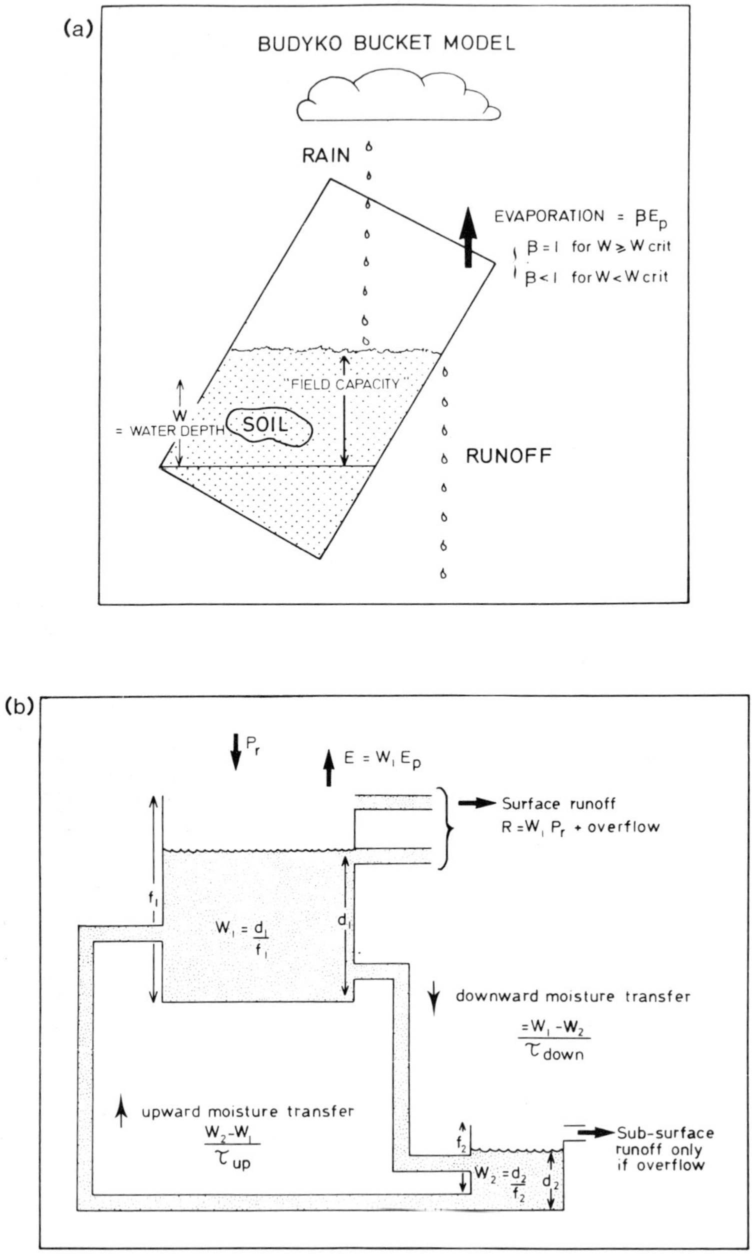

Figure 6.13 (a) The Budyko bucket model land-surface parameterization; (b) processes involved in a two-soil-layer land-surface scheme. The relative saturation of the soil columns is often represented by W = depth/field capacity. The evaporative loss is generally a fraction of the potential evapotranspiration, E_p. (Note the absence of vegetation in both the land-surface schemes shown in this figure).

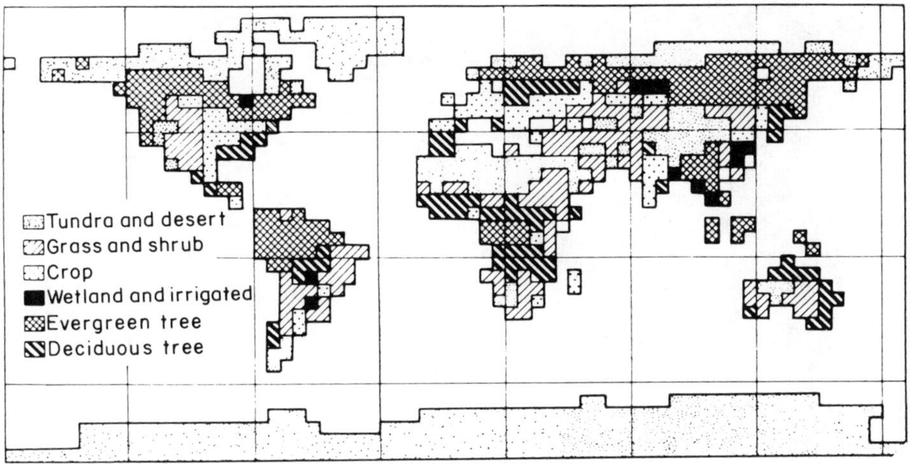

Figure 6.14 Distribution of generalized ecotypes on a $5° \times 5°$ grid. The coarse spatial resolution of both geography and the simple ecological classification required by most GCMs is difficult to derive from either atlas or satellite data.

features. In some senses the data source is not the paramount problem but rather how averaging or agglomeration can be conducted so that the land-surface characteristics can be described at a resolution suitable for input to general circulation climate models. Figure 6.14 shows an example of a classification of major ecotypes grouped into six classes and archived on a $5° \times 5°$ grid which is approximately equivalent to the grid used by most GCMs. The coarseness of both the geography and the ecotype classification is immediately apparent.

The situation is significantly worse for other land-surface information. Only one hydrographic data archive and a couple of soils data sets are currently available. Figure 6.15 is an illustration of both one relatively gross classification of soil colour and also the impact of producing an archive at a much finer resolution ($1° \times 1°$ in this instance). These archives differ in resolution, complexity, production methodology, reliability (Figure 6.16) and in character. The difficulty of modelling the interaction between vegetation and surface hydrology is exacerbated by the requirement that parameterization schemes must be appropriate to GCM grid squares of $\sim 5° \times 5°$. Many botanists despair when they are asked questions such as, ‘What is the root resistance for a particular vegetation type characterized on an eight-point scale which describes all the world’s vegetation?’ It is indeed hard to understand what such a root resistance means in the context of a GCM and, more importantly, how it can be related to measurable quantities. Other components of the GCM may well be inadequate or unsuitable. For example, one modelling group has had to develop a new diurnal cycle planetary boundary layer treatment for their GCM before they could incorporate a more sophisticated surface scheme.

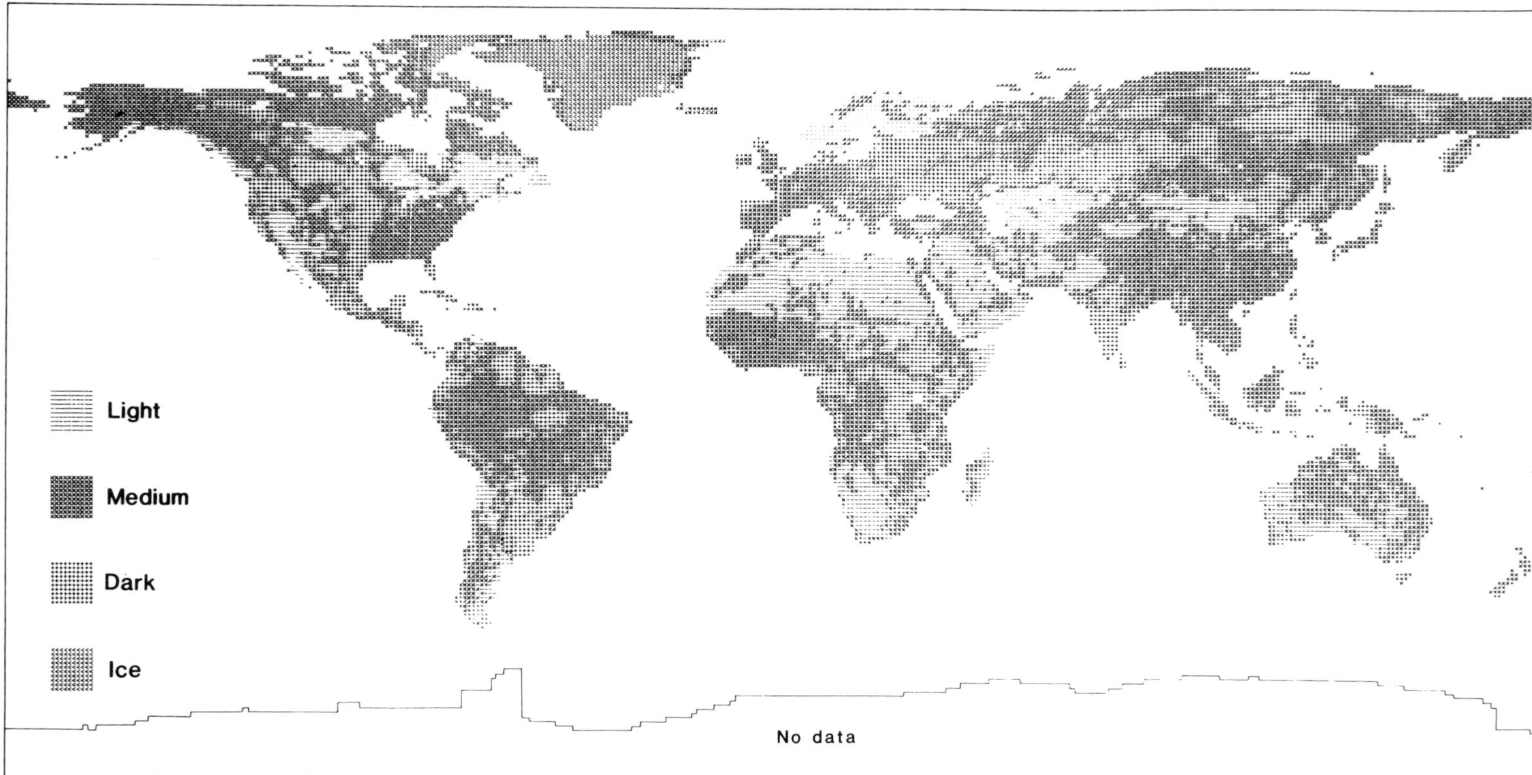

Figure 6.15 Global distribution of soil cover on a $1° \times 1°$ grid. Soils have been graded on a three-point scale (after Wilson and Henderson-Sellers, 1985).

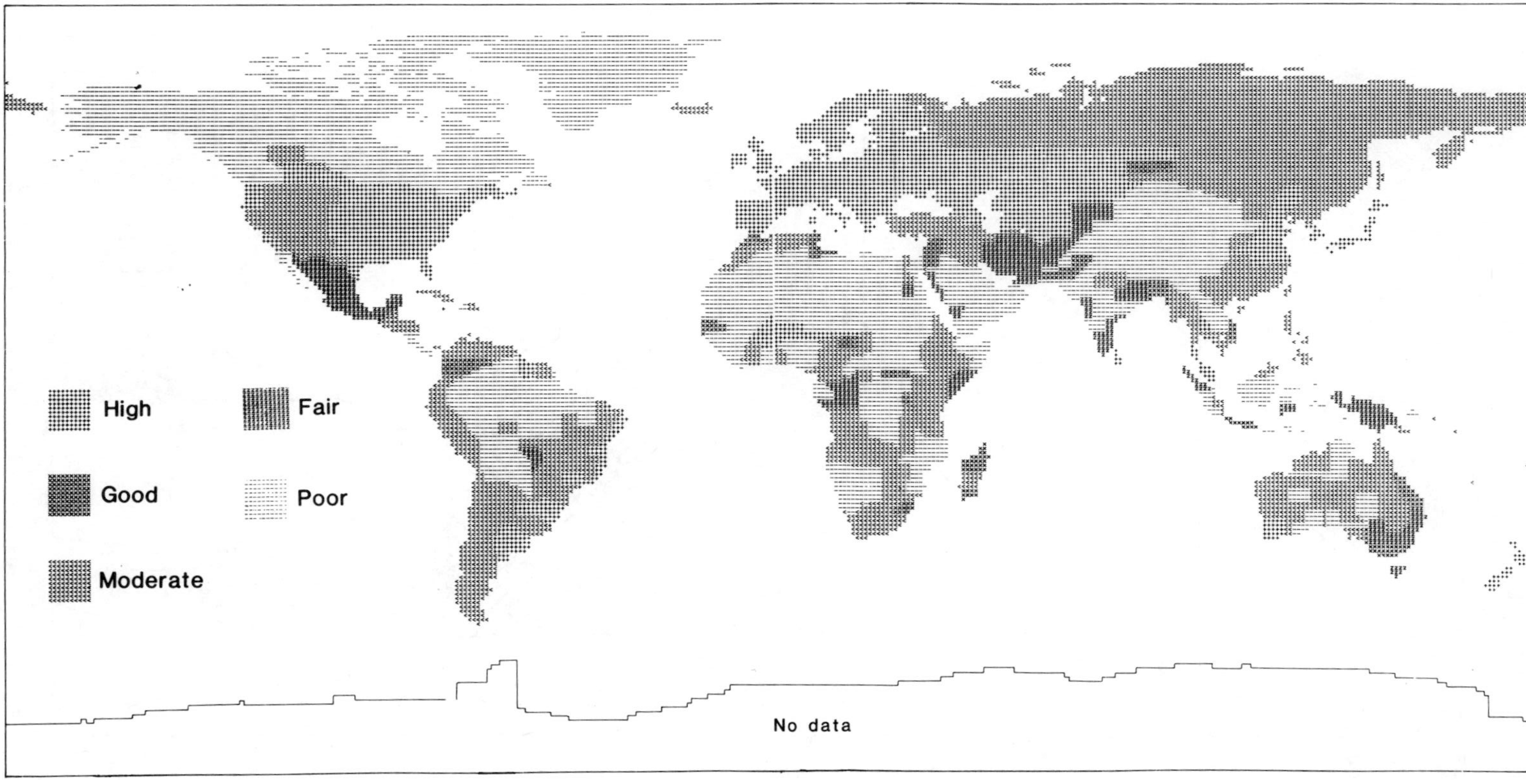

Figure 6.16 Reliability estimates ($1° \times 1°$) for the source data used in compiling the soils data set shown in Figure 6.15 (after Wilson and Henderson-Sellers, 1985).

(a)

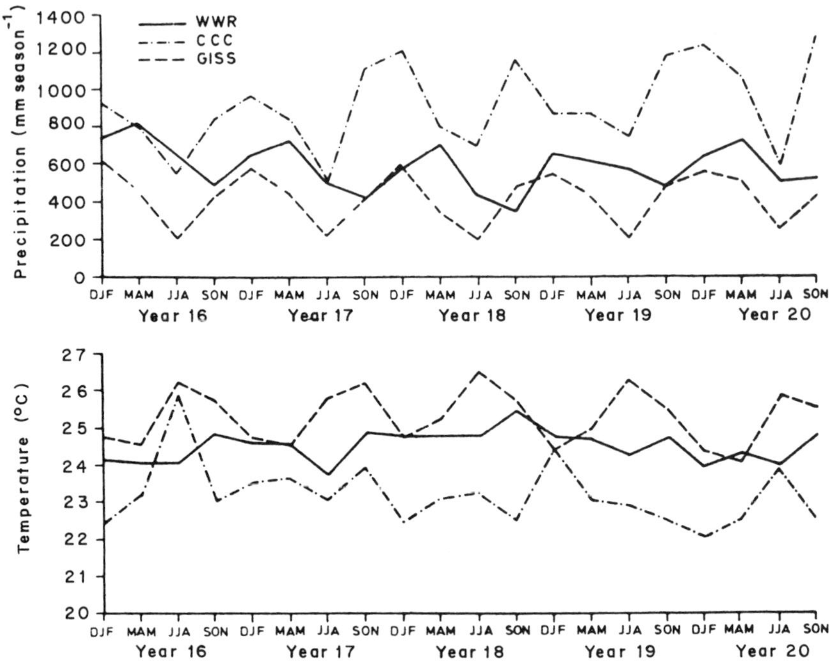
WWR
CCC
GISS
Precipitation (mm season⁻¹)
1400
1200
1000
800
600
400
200
0
DJF MAM JJA SON DJF MAM JJA SON DJF MAM JJA SON DJF MAM JJA SON DJF MAM JJA SON
Year 16
Year 17
Year 18
Year 19
Year 20
Temperature (°C)
27
26
25
24
23
22
21
20
DJF MAM JJA SON DJF MAM JJA SON DJF MAM JJA SON DJF MAM JJA SON DJF MAM JJA SON
Year 16
Year 17
Year 18
Year 19
Year 20

(b)

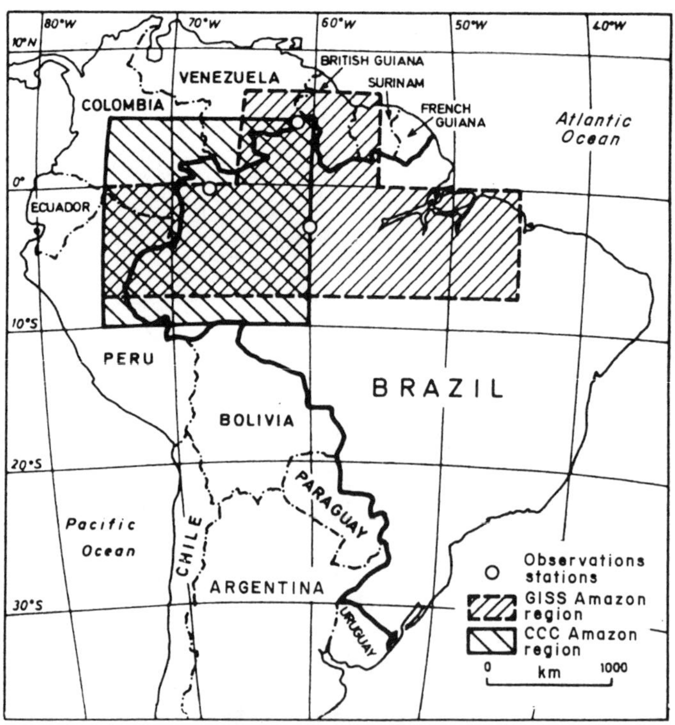
80°W
70°W
60°W
50°W
40°W
10°N
0°
10°S
20°S
30°S
VENEZUELA
BRITISH GUIANA
SURINAM
FRENCH GUIANA
COLOMBIA
Atlantic Ocean
ECUADOR
PERU
BRAZIL
BOLIVIA
PARAGUAY
Pacific Ocean
CHILE
ARGENTINA
URUGUAY
Observations stations
GISS Amazon region
CCC Amazon region
0
km
1000

Application to the tropical Amazon

Figure 6.17(a) shows the results from the last 5 years of long-integration control runs of the GISS and the Canadian Climate Centre (CCC) GCMs. The plots are of cumulative seasonal precipitation and seasonally averaged temperature for the 'Amazonian' regions of the two models. These regions differ somewhat as is shown in Figure 6.17(b). The CCC GCM is a spectral model truncated at wavenumber 20 (a spatial resolution of $\sim 5°$) and the Amazonian results are drawn from nine grid elements which cover the upper part of the Amazon Basin. The GISS GCM is a coarse-resolution cartesian grid model (grid elements have dimensions 8° by 10°). In this model the Amazonian region is represented by four grid elements spanning most of the region of Legal Amazonia (Figure 6.17(b)). Also shown in Figure 6.17(a) are plots of observed temperature and rainfall averaged from data for stations selected from the World Weather Record and lying within both GCM Amazonian regions. For both GCMs equilibrium has presumably been achieved, since the results shown are for the last 5 years of 20-year simulations.

The seasonality in the precipitation regime is stronger in both GCMs than in the composite observational data. Rainfall data do show a strong seasonality but it is out of phase between the northern and southern Amazon. The fact that both models exhibit the seasonality of the southern Amazon suggests that modelled rain belts may be displaced northwards, especially in the CCC GCM. It is important not to draw any conclusions about the verity of the climatic simulations with these two models, particularly as it must be noted in the context of Figure 6.17(a) that the observational data are at least as suspect as the results of the GCMs. The results do, however, illustrate the primary difficulty inherent in evaluating the performance of GCMs: in specific locations for which observational data can be ascertained slight shifts in the model circulation combine with the coarse resolution to make validation very difficult.

(Caption to Figure 6.17 *opposite*)
Figure 6.17 (a) Seasonal precipitation (mm season^{-1}) (upper) and average seasonal surface temperature (°C) (lower) as computed in the CCC and GISS GCMs for the 'Amazonian' regions shown in (b). The observational results have been computed from data in the World Weather Record (WWR) for the years 1951–60. Values are averages for three stations: Manaus (03°08′S; 60°00′W); Uaupes (00°08′S; 67°05′W); Santa Elena (04°36′N; 61°07′W) within the 'Amazonian' areas of the two models. (b) Two grid areas from which the simulation results shown in (a) were drawn. The CCC GCM area is composed of nine $\sim 5° \times 5°$ grid elements while the GISS GCM area is composed of four $8° \times 10°$ grid elements. The spatial resolution and the areal coverage of these 'Amazonian' regions differ (after Henderson-Sellers, 1987).

The simplest way to undertake basic sensitivity studies of the parameters and the parameterizations of a land surface scheme is to operate the scheme as a stand-alone zero-dimensional model since perturbed and control simulations can be compared absolutely, without any need to refer to statistics. The land-surface portion of a version of the NCAR Community Climate Model (CCM) is more complex than many schemes incorporated into other GCMs, although it is still highly simplified compared with reality (Figure 6.18). It incorporates most of the essential surface features, including a vegetation canopy, surface and rooting zone soil layers, variable albedo and hydrological characteristics. The soil model consists of two layers—an upper soil layer interacting directly with the surface air and embedded in a total soil column acting as a deeper 'store' for moisture and energy. Depths of both soil layers are variable, as is the proportion of the plant roots, distributed within the upper rather than the lower layer. Thus, it is possible to differentiate between predominantly surface-rooting plants and those which 'tap' soil moisture held at depth. Energy and moisture fluxes are calculated as a function of the soil type which controls texture (and hence porosity, hydraulic conductivity, etc.) and colour (also a function of soil wetness). However, both latent and sensible fluxes depend on the vegetation canopy properties and the plant canopy is represented by a single layer.

Using the model in its stand-alone mode air temperature must be specified, for example varying sinusoidally throughout the day. The net surface infrared radiation is computed using canopy and ground temperatures for the upward flux and calculating an effective atmospheric emissivity from the specific vapour pressure to estimate a downward flux. Incident solar radiation, relative humidity, wind speed and precipitation also need to be specified, and during precipitation events the incident solar radiation could be, for example, significantly reduced to simulate the effects of increased cloudiness. The saturated specific humidity of the air is calculated as a function of the specified air temperature at each timestep. Some examples of the types of results which can be achieved are described below.

Changes in the leaf area index (LAI) alter the ratio of transpiring to non-transpiring canopy surface and the area over which dew formation and interception occur. More water can be retained in the case of a higher LAI. The total surface resistance to transpiration is the resistance per unit leaf area divided by the LAI. Hence, average canopy stomatal resistance is smaller with a larger LAI. The greatest response to perturbation of LAI should be in the leaf energy and moisture budget and in the surface moisture regime. These responses can be illustrated in an example integration representing a low-latitude evergreen forest (Figure 6.19(a)). The largest response is found to be in the leaf temperature. For a LAI reduction experiment the leaf temperature increases by ~4 K. A number of different effects resulting from the reduction in LAI contribute to this temperature change. Less moisture is intercepted, causing a reduction in the flux of latent heat directly from the vegetation surface. The

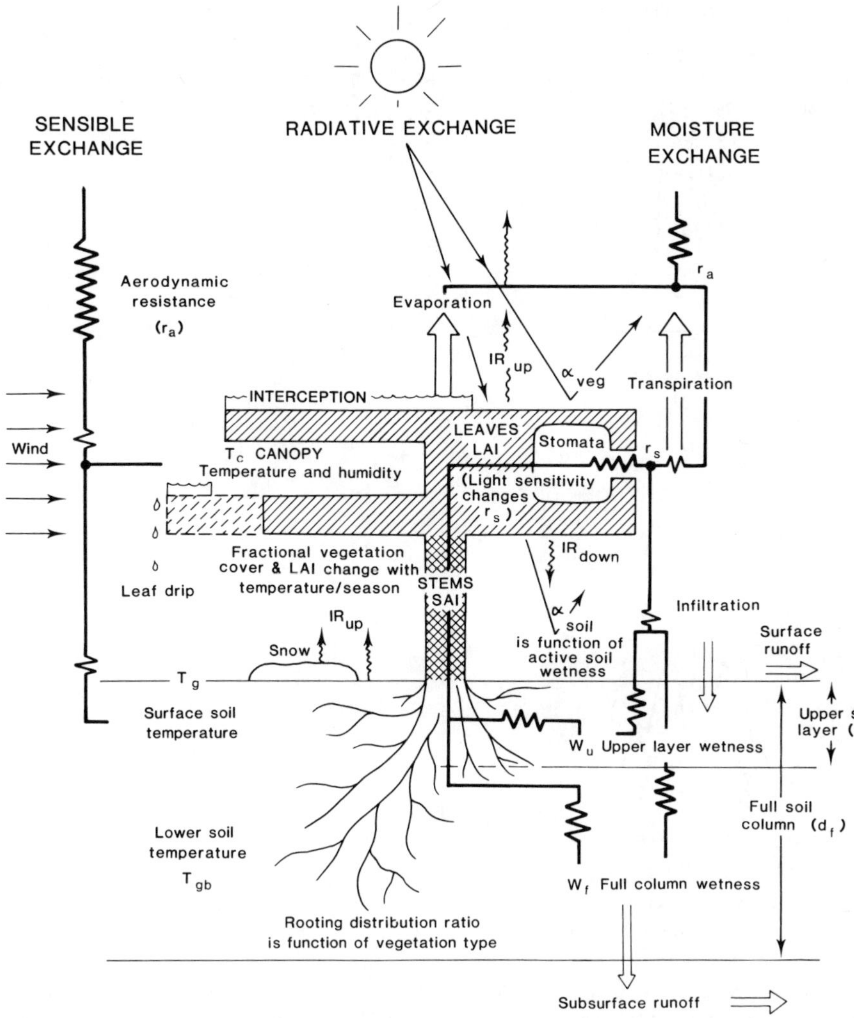

Figure 6.18 Schematic representation of the maximum level of complexity so far achieved in land-surface parameterization schemes in GCMs. This scheme controls the radiative, latent and sensible fluxes occurring at the surface–atmosphere interface. Note especially that the 'plant' almost covers one grid element, i.e. $\sim 5° \times 5°$, or approximately 500 km × 500 km.

total vegetation area is reduced, as is the turbulent exchange between the air and the canopy. Evapotranspiration decreases by up to 100 W m^{-2} throughout much of the day, although evaporation from the soil is greater. More water enters the soil and soil moisture is greater. The pattern of response is reversed and the soil moisture is less when the maximum LAI is increased.

In the case of a different land surface type, tundra, in the spring with the snow melting, the tundra leaf temperatures differ little between the two perturbed cases of increased and decreased LAI. However, although the runoff is very

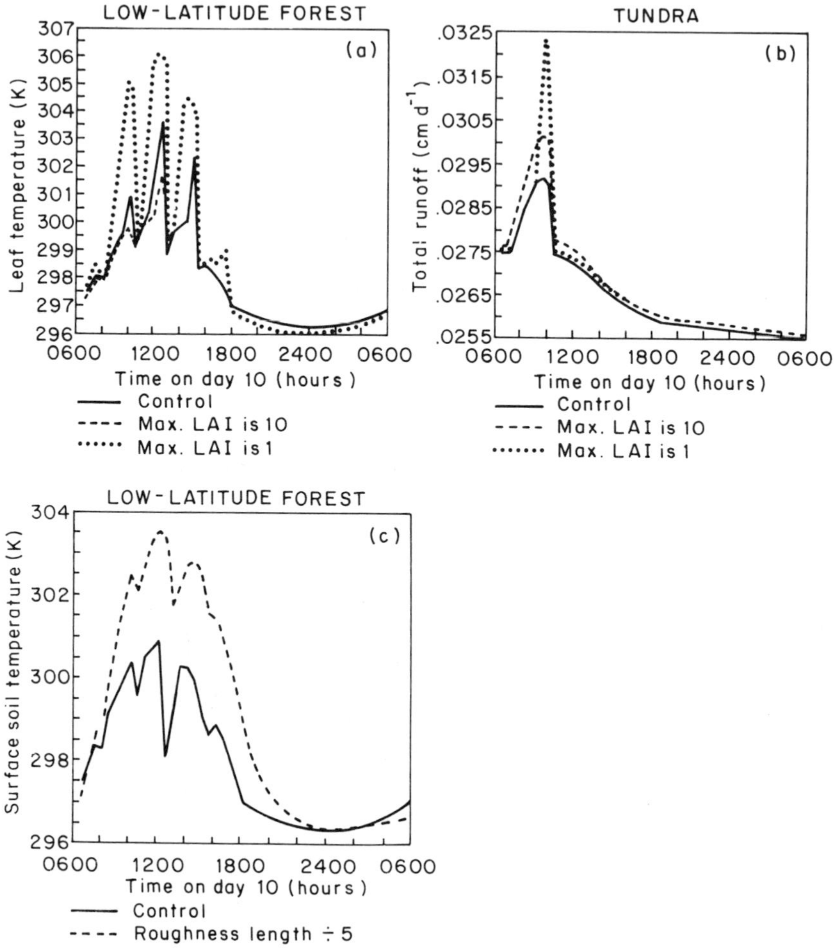

Figure 6.19 (a) Difference in leaf temperature for different values of LAI for a low-latitude forest; (b) differences in total runoff for an experiment on the LAI of tundra; (c) differences in soil temperature for a change in roughness length of a low-latitude forest. All of these results have been generated using the land surface scheme shown schematically in Figure 6.18 operated in a stand-alone mode in order to establish the land-surface sensitivity to prescribed parameter values.

small (Figure 6.19(b)), there are notable differences between the two extreme values of LAI. In particular, the total runoff is sharper in time and attains a higher value in the case of LAI reduced to 1 as a result of leaf moisture attaining its prescribed maximum depth of 0.02 cm earlier.

As the roughness length increases, the bulk neutral transfer coefficient (cf. equation (6.14)) increases and vice versa. For equivalent surface-atmosphere temperature and moisture gradients, heat and moisture fluxes will be greater for a larger roughness length. Decreasing the roughness length of the low-latitude forest (Figure 6.19(c)) reduces the energy and moisture fluxes. Increased soil moisture prolongs potential evaporation from wet leaves and, in this case, increases soil temperatures. The low-latitude forest responds similarly. When its roughness length is increased, the response is influenced by strong feedback between soil moisture and energy fluxes. With increased roughness, most of the rainfall on the second day of the simulation is evaporated without entering the soil. Consequently, in spite of an incrcase in the surface bulk transfer coefficient, increased surface resistance during the drying period reduces the evapotranspiration. (It is interesting to compare the complex responses of this sub-scheme with the parameterization of the biosphere in the first of the simple climate models described in Appendix B.)

Observational data are urgently needed to validate land surface parameterization schemes such as this which are designed to simulate surface properties and responses areally averaged over regions of size $\sim 4.5° \times 7.5°$. Current data are inadequate although regional experiments within the International Satellite Land Surface Climatology Project (see Section 6.7) may eventually produce useful data.

6.6 COUPLED OCEAN–ATMOSPHERE GENERAL CIRCULATION CLIMATE MODELS

One important and recent development in modelling the climate system is the coupling of ocean and atmospheric GCMs. Until very recently the oceanic component of the climate system has been specified either by the prescription of fixed (seasonally varying but present-day) sea-surface temperatures or by prescribing the meridional energy transport of the oceans. This clearly constrains the simulation and many authors have identified the urgent need to include the ocean subsystem with the deep ocean as well as the upper mixed layer in three-dimensional climate models.

The difficulties inherent in ocean–atmosphere coupling are identified in Figure 6.20. This diagram underlines the different response times associated with the atmosphere and the ocean subsystems and emphasizes that the ocean subsystem spans a greater range in *both* time and space than the atmosphere subsystem. Thus a sophisticated ocean GCM which has eighteen levels in the vertical and 30 m resolution spatially is barely able to capture the eddy features vital for ocean dynamics. Topography is as important for ocean modelling as for atmospheric modelling but basin geography is much more important than bottom topography. The considerable discrepancy in response (or equilibration times) of the atmosphere and the ocean including the deep ocean have already been described (see Section 1.3 and especially Table 1.1). Since the time to reach equilibrium is much longer for the ocean than for the atmosphere, linking of an ocean model with an atmospheric model is very difficult (Figure 6.21). One modelling group found it suitable to synchronize 4.2 years of atmospheric model integration with 1200 years of ocean model integration.

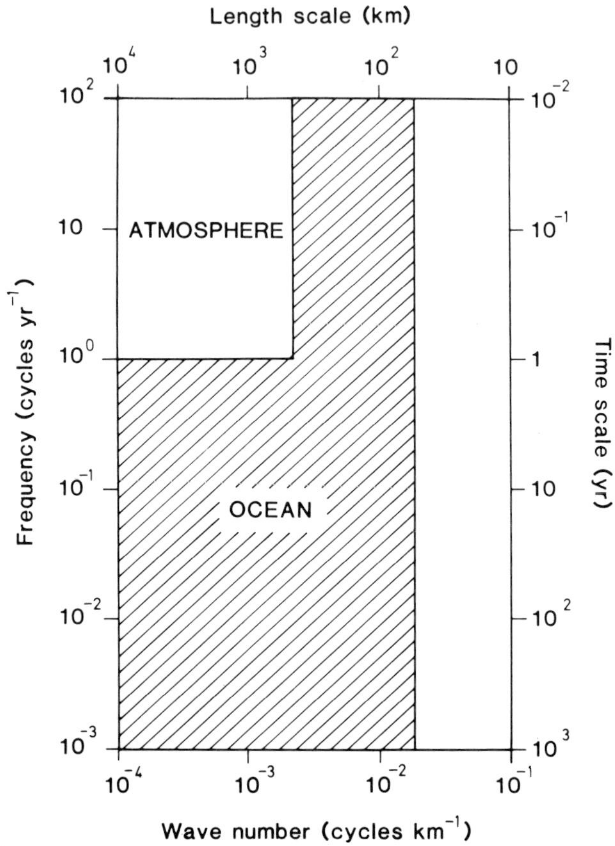

Figure 6.20 The range of significant horizontal wavenumbers and frequencies in climate models of the atmosphere and ocean. The lowest frequency is determined by the thermal relaxation times of the subsystem and the highest by gravity waves. The lowest wavenumber is determined by the planetary scale and the highest by the radius of deformation of the subsystem medium. The inverse of the scales gives the more familiar time and length scales. Note that the ocean spans a greater range of *both* fundamental scales than the atmosphere.

Ideally the linkages should be between the thermodynamic systems, among the variables represented in the equations of motion and in terms of the parameters and variables of the water cycle (Figure 6.21(b)). Close interaction between the ocean sub-model and the atmospheric sub-model is impossible because of the requirement that the ocean subsystem be allowed a much longer time to equilibrate. The procedure which is adopted is to use a jointly accessible library within the computer program. Annual and monthly mean values of all the variables of mutual interest are stored and updated in this library. They are therefore jointly accessible. The library elements are updated as the oceanic part of the model is integrated for 1200 annual cycles, while remaining accessible to the atmospheric sub-model. Hence rates of exchange of heat, momentum and moisture

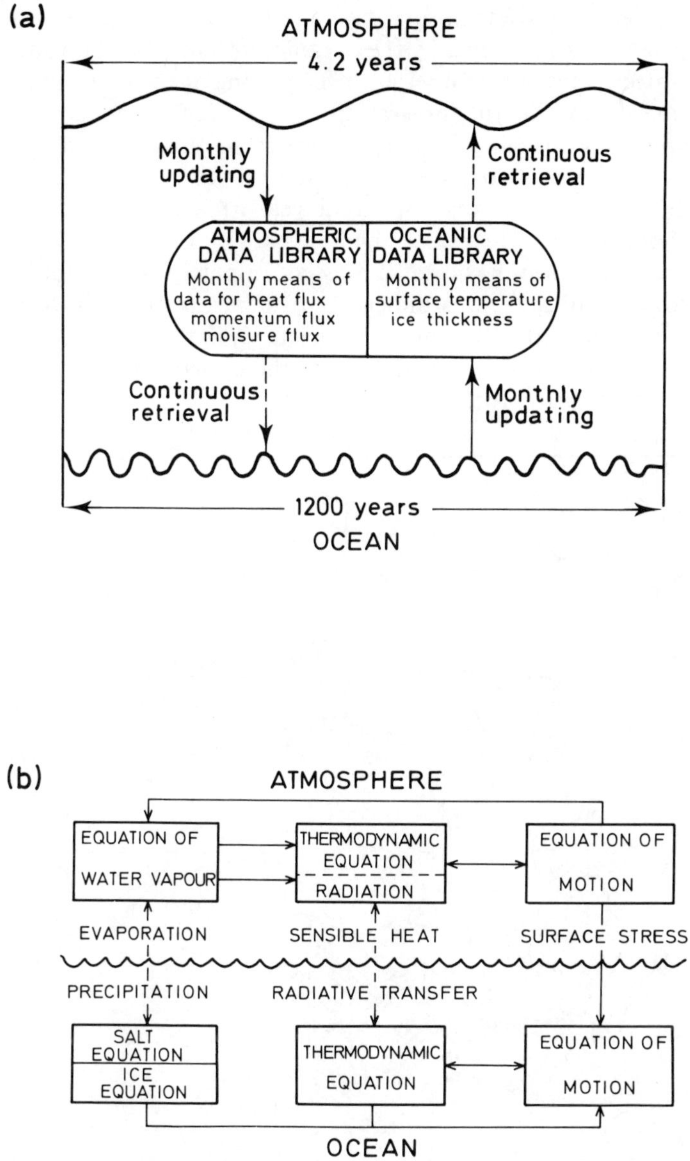

Figure 6.21 (a) Diagram of the coupling procedure used for a joint ocean–atmosphere model with seasonal variation of insolation; (b) box diagram illustrating the major components of a joint ocean–atmosphere model and the interaction among the components (both reproduced with permission from Manabe *et al.*, 1979).

can be computed from these library data and the latest values of atmospheric variables. The fluxes computed by the oceanic sub-model and the atmospheric sub-model differ from one another at the beginning of the time integration but approach one another as the integration of the coupled model progresses.

While many early mixed layer models investigated mixed layer deepening over relatively short timescales (frequently less than 6 months) it has long been realized that, for GCM applications, an inherent year-to-year numerical stability is necessary. Some concern regarding the applicability of mixed layer models to longer period simulations has been expressed. In particular a detailed study of upwelling/downwelling regions undertaken by means of the inclusion of the wind stress led to identification of the deficiencies in the approach with regard to the role of surface heating in restabilizing the upper mixed layer, suggesting it to be an area ripe for further developments.

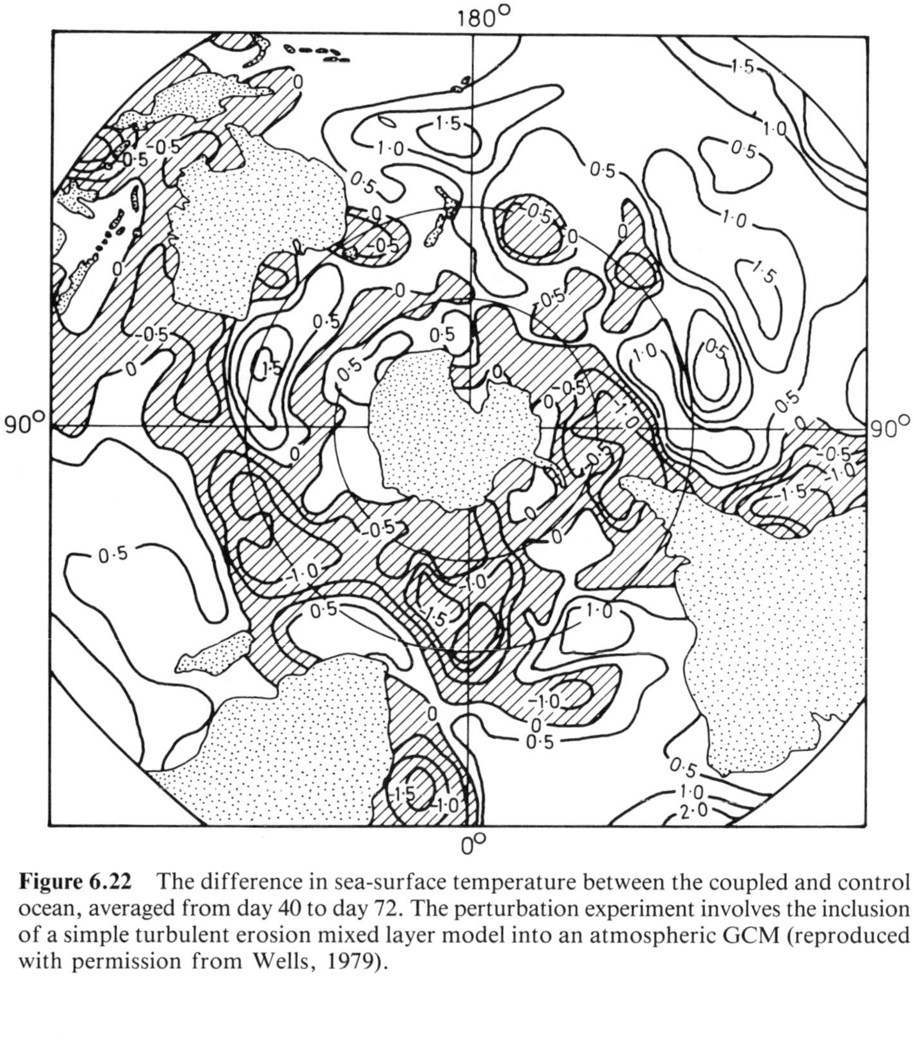

Figure 6.22 The difference in sea-surface temperature between the coupled and control ocean, averaged from day 40 to day 72. The perturbation experiment involves the inclusion of a simple turbulent erosion mixed layer model into an atmospheric GCM (reproduced with permission from Wells, 1979).

It has been suggested that modified turbulent erosion models are appropriate for middle latitudes since the sea-surface temperatures are governed largely by the local exchanges of heat and mechanical energy in these latitudes; whereas in equatorial latitudes, lateral advection of energy needs to be included. The results of coupling a simple turbulent erosion model to a southern hemisphere spectral GCM are shown in Figure 6.22. Although advection is included within the mixed layer it is neglected below this depth. However, no simultaneous coupling of the ocean–atmosphere system was used. Instead the atmospheric model was integrated for 4 days with fixed sea-surface temperatures, which were then themselves recalculated every fourth day. Results from this model were compared with a simulation in which the atmosphere forces the ocean but no feedback of sea-surface temperatures on the atmosphere was included. Conclusions regarding the differences between these two model runs were limited to simulation periods of 72 days, despite the fact that simulation spin-up periods are usually considered to be years for the atmosphere and centuries for the ocean. One further problem is that of incompatible time-scales: typically a mixed layer model uses a timestep of 1 hour as compared to the advective component of an oceanic GCM with timesteps of several hours.

Parallel development of ocean circulation models and mixed layer models has led to the need to embed the latter within the former to produce an efficient three-dimensional ocean circulation model. Recently, the coupling of ocean–atmosphere models has been described in terms of a hierarchy of oceanic components. It is suggested that even oversimplistic oceanic representations, e.g. (i) the 'swamp' model with no heat storage capacity and (ii) fixed depth ocean surface layer models where there is a heat capacity but no dynamics, may be useful in helping to understand the processes acting in air–sea interaction. The advantage of utilizing a prescribed depth mixed layer in permitting the inclusion of a full seasonal cycle in the atmospheric GCM (not possible with the oceanic 'swamp') is considerable. It is possible to utilize a model in which oceanic mixed layer depth is allowed to vary in a prespecified way, while acknowledging that more complete (and hence more complex) models of oceanic dynamics and thermodynamics exist—either in terms of a full computation of the oceanic surface layer or in terms of a three-dimensional oceanic circulation model.

The use of prespecified mixed layer depth and calculated sea-surface temperatures is inadequate for the simulation of the annual cycle of zonal heat storage, especially in the tropics. Preliminary results have been presented in the literature which compare three oceanic representations: a simple diffusive convective two-layer model; a two-layer model, but including a prognostic model of the mixed layer; and a three-layer primitive equation model. The results of this study were that within this hierarchy, accuracy and usefulness increase with increasing complexity of oceanic mixed layer model, but so does the computational time required.

Table 6.3 Data specification for the International Satellite Cloud Climatology Project (reproduced by permission of the American Meteorological Society from Rossow *et al.*, (1985), *Journal of Climate and Applied Meteorology*, **24**, 887–903.

	Precision (30-day averages)
Amounts	
Total cloud amount (fraction)[a]	±0.03
Cirrus cloud amount (fraction)[a]	±0.05
Middle cloud amount (fraction)	±0.05
Low cloud amount (fraction)[a]	±0.05
Deep convective cloud amount (fraction)	±0.05
Height	
Cirrus cloud-top height (km)[a]	±1.00
Middle-level cloud-top height (km)	±1.00
Low-level cloud-top height (km)	±0.50
Deep convective cloud-top height (km)	±1.00
Cloud-top temperature (K) for each cloud category[a]	±1.00
Cloud optical depth	
Cloud size distribution	
Average narrow band radiances (visible and infrared)[a]	
Spatial averaging—The information is to be averaged over approximately 250 × 250 km boxes	
Time sampling—Every 3 hours, i.e. eight times a day, centred on the synoptic observation times	
Time averaging—The global cloud climatology will consist of 30-day averages for each of the eight observing times per day	
Length of time series—5 years	
Parameters—Spatial and temporal averages and variances (or another statistical measure of the shape of the temporal distribution) are required for each of the parameters	

[a] Highest priority.

6.7 FUTURE CLIMATE PROJECTS AND THEIR IMPORTANCE TO GENERAL CIRCULATION CLIMATE MODELS

There is currently a wide range of investigative projects in progress which are designed to aid understanding of the climate system and to provide data for validation and improved parameterization of global climate models.

Observational programmes

The monsoon experiment (MONEX) was the principal field experiment of the monsoon sub-programme of the First GARP Global Experiment (FGGE). The study, over the Arabian Sea, Indian Ocean, Bay of Bengal and Southeast Asia provided a great deal of information on the behaviour of the intertropical convergence zone and also on larger scale events in both hemispheres as well as on the Indian subcontinent monsoon itself.

The Tropical Ocean Global Atmosphere (TOGA) project formally began in January 1985. It is a 10-year project including measurements of a wide range of variables pertaining to the atmosphere, ocean and the ocean–atmosphere interface. In the latter half of the 10-year programme, ocean-orientated satellites, still to be launched, will augment the surface-based observational network. The observational domains for the TOGA programme are the tropical oceans from about 20° N to 20° S and the global atmosphere. One of the unique aspects of TOGA is its aim of measuring the fluxes between these two domains. When complete, TOGA will provide vital data for the improvement of coupled ocean–atmosphere models. In contrast to TOGA, WOCE (the World Ocean Circulation Experiment) will place emphasis on determining the structure and evolution of the middle and deep ocean. Measurements will be largely from surface ships but satellite altimeter data will again play an important role.

The International Satellite Cloud Climatology Project (ISCCP) began formally on 1 July 1983. Its aim is to produce a 5-year, homogeneous data archive of the world's cloud cover as viewed from satellites. The particular characteristics which scientists involved in ISCCP aim to derive are listed in Table 6.3. The timing of ISCCP was designed to coincide with the availability of satellite data from five geostationary and two polar-orbiting satellites. This satellite array is likely to be the best data source from the point of view of cloud retrieval for some time in the future.

The land surface is being investigated under two projects organized through the World Meteorological Organization. The International Satellite Land Surface Climatology Project (ISLSCP) was begun in 1985 and aims to conduct a series of experiments culminating in the early 1990s in a closely interlocked set of measurement programmes ranging in size from satellite retrieval of land-surface features at a scale of tens of metres up to measurement on scales equivalent to the gridscale of atmospheric GCMs (Figure 6.23). The Hydrologic–Atmospheric Pilot Experiment (HAPEX) is a less ambitious scheme designed to monitor hydrological properties of a 100×100 km area in southwest France beginning in the year 1986 (Figure 6.24). The data derived from HAPEX will be of direct use for the improvement of land-surface parameterization schemes such as those described in Section 6.5 and the results will also link in directly to the ISLSCP.

There are many other regional observational and regional modelling projects: for example, ALPEX is designed to try to understand the effect of local-scale topography on regional meteorological and climatological patterns in the immediate area of the Alps. Results from ALPEX have nurtured improvements in the prescription of topography in some medium-range forecast models.

Questions for the future

Another important area currently under investigation is the best method of assessing climate sensitivity objectively. The term 'climate sensitivity' has two meanings. The first, which has been stressed throughout this book, is when the term is used to indicate the response of a model to changes either in prescribed

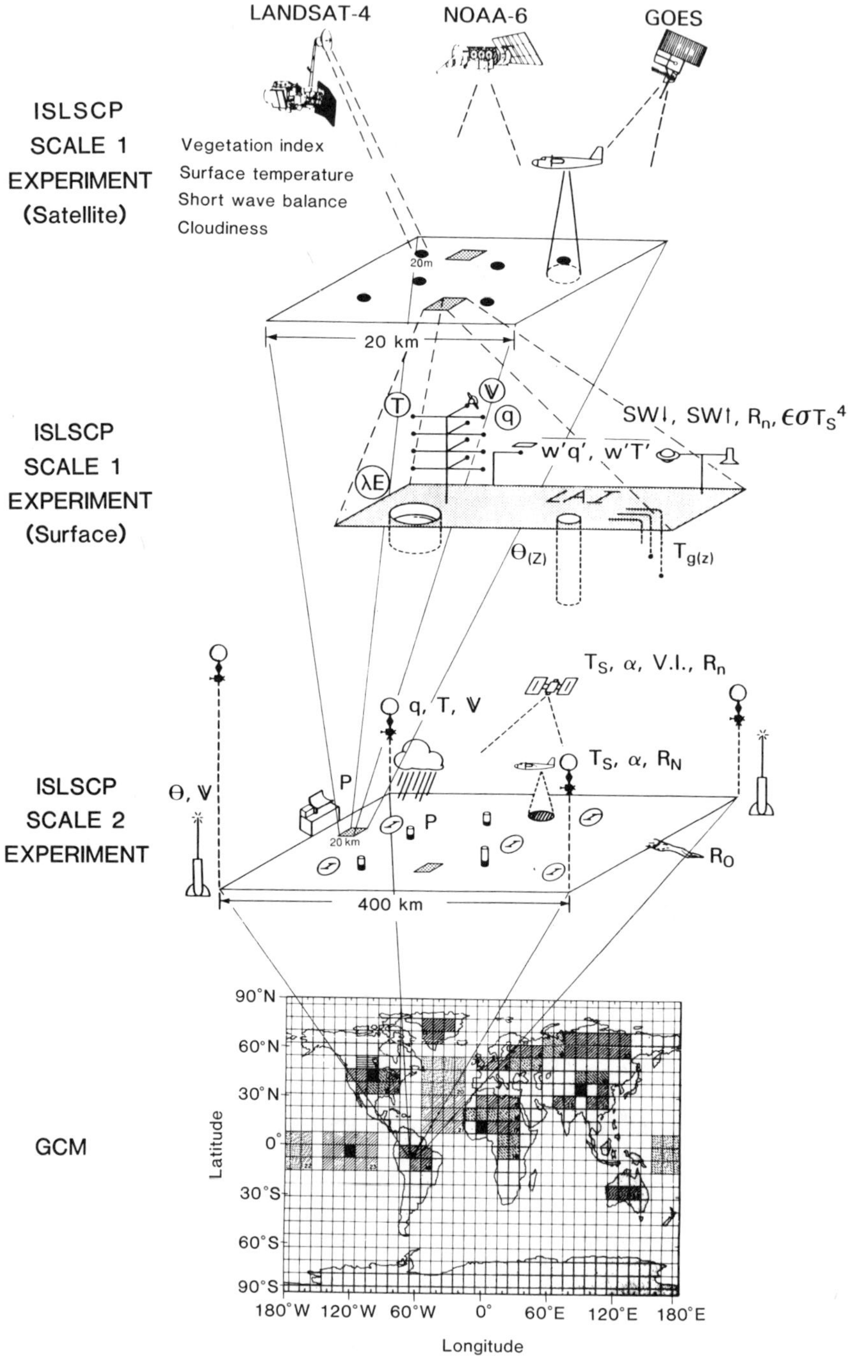

LANDSAT-4
NOAA-6
GOES
ISLSCP
SCALE 1
EXPERIMENT
(Satellite)
Vegetation index
Surface temperature
Short wave balance
Cloudiness
20m
20 km
ISLSCP
SCALE 1
EXPERIMENT
(Surface)
SW↓, SW↑, R_n, $\epsilon\sigma T_S^4$
$\overline{w'q'}$, $\overline{w'T'}$
$\Theta_{(Z)}$
$T_{g(z)}$
ISLSCP
SCALE 2
EXPERIMENT
Θ, V
q, T, V
T_S, α, V.I., R_n
T_S, α, R_N
P
P
20 km
R_O
400 km
GCM
90°N
60°N
30°N
0°
30°S
60°S
90°S
Latitude
180°W 120°W 60°W 0° 60°E 120°E 180°E
Longitude

external parameters or in model constants and parameterizations. The second, and arguably the more important, is its use as a description of the potential response of the climate system to the effects of mankind.

We can use a climate model to determine whether some perturbation has had a significant effect on the climate. In simple models such a question is easily answered. However, in more complex models there is an element of randomness so that eventually a model 'loses its memory' of its initial conditions. Consequently the question of statistical significance arises. The simplest approach to examining a model response is to consider how the signal compares with the variance of both the control and perturbation experiments.

The situation is not as simple as it might first appear. While this technique would work for just surface temperature, in cases where there are many variables being modelled and interacting, then a significant change in one variable may simply be due to a random superposition of random changes in other model variables. It is perhaps important, therefore, to take great care in the use of statistical confidence limits in reporting climate-modelling studies. Consequently, interpretation of model results must involve some degree of physical analysis of mechanisms and of risk evaluation. The appropriate statistical concept is therefore decision-making. We need to know how large the perturbation must be before the effects become significant and indeed what we mean by 'significant', both statistically and within the framework of societal impact.

The temptation to try to make a prediction before the model is tested and demonstrably viable is great. Climate modelling is a very young science in which discoveries occur often and not always how and where they might be anticipated. Figure 6.25, for example, shows a pair of plots of ocean deep water and surface water temperatures computed using a coupled ocean–atmosphere GCM for a range of perturbations to the level of atmospheric CO_2. This response can be compared with temperatures predicted by the simple ocean plus atmosphere box model described in Appendix B. The results shown in Figure 6.25 are, naturally, of interest in the context of 'the impact of increasing CO_2' but also reveal an unexpected 'application'. In Section 1.6 (see Figure 1.16) the question of the anomalously high bottom water temperatures in the Cretaceous was posed. It is tantalizing to view the results for the low-latitude case in Figure 6.25 in the light of this puzzle. The model suggests that increasing atmospheric CO_2 fourfold produces temperature differences (from the present day) of the magnitude seen in the observational record. As climate modelling becomes more sophisticated it is important that we do not lose sight of unexpected possible connections such as this.

(Caption to Figure 6.23 *opposite*)

Figure 6.23 The variety of scales of observation envisaged as forming integral parts of the International Satellite Land Surface Climatology Project (ISLSCP). Measurement at all of these scales of resolution is required in order to determine land surface characteristics for GCM studies.

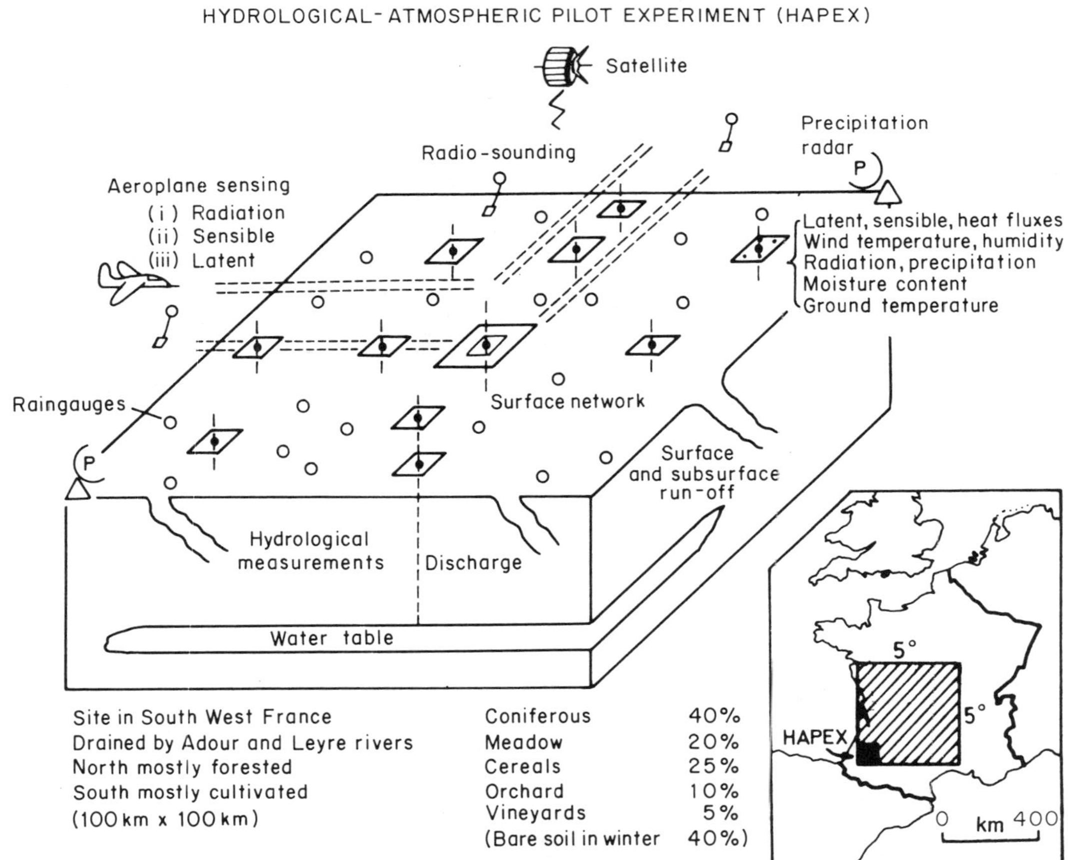

Figure 6.24 Diagrammatic representation of the surface and atmospheric measurement systems used during HAPEX in 1986. The particular emphasis of this regional project was the measurement of land-surface hydrology with a view to improving land-surface parameterizations in GCMs.

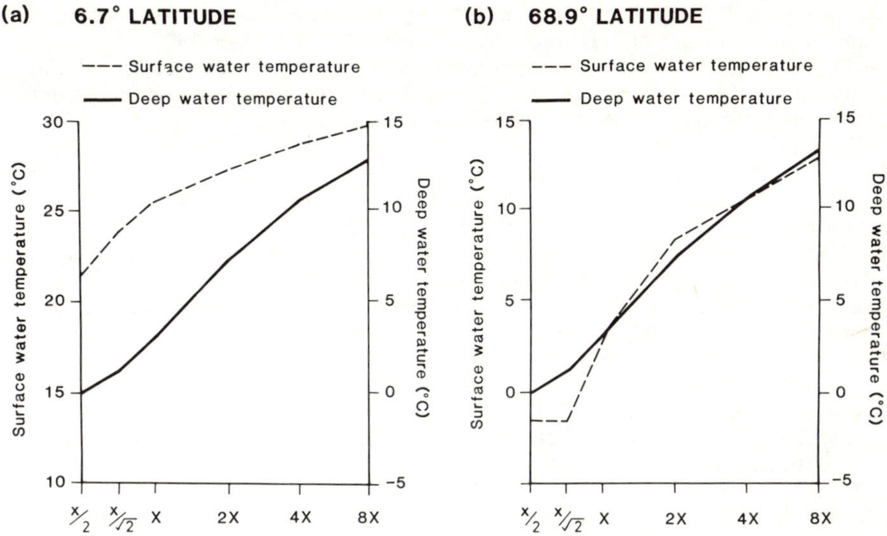

Figure 6.25 Zonal mean temperatures of surface and deep water in the equilibrium solutions at 6.7° and 68.9° latitude for a range of atmospheric CO_2 concentrations (present-day $= X$) as predicted by a coupled ocean–atmosphere model (from Manabe and Bryan, *J. Geophys. Res.*, **90**, 11689–11707 (1985), Copyright by the American Geophysical Union).

6.8 EPILOGUE

This book has been written with the basic aim of building up a framework within which all types of climate models can be considered. It has been our intention to stress two themes in particular: (i) that the time is now right for the recruitment of a wide variety of scientists, social and economic researchers and political and demographic planners into the blossoming field of climate modelling, and (ii) that in the focusing on application and prediction which will dominate a great deal of thinking in the immediate future we, as a community, avoid becoming blinkered.

We have repeatedly drawn readers' attention to unexpected results of climate modelling and to the importance of using simple models as well as very complex ones, particularly for identifying mechanisms and to aid one's own understanding. The power and range of application of climate models are very great. Modellers cannot yet hope to offer complete answers to the important questions pertaining to man's environment, but the continuing search for a more complete understanding of the climate system is a most laudable endeavour.

RECOMMENDED READING

Boer, G. J., McFarlane, N. A., Laprise, R., Henderson, J. D., and Blanchet, J-P. (1984) The Canadian Climate Centre spectral atmospheric general circulation model, *Atmos-Ocean*, **22**, 397–429.

Deardorff, J. (1978) Efficient prediction of ground temperature and moisture with inclusion of a layer of vegetation, *J. Geophys. Res.*, **83**, 1889–1903.

Dickinson, R. E. (1984) Modelling evapotranspiration for three dimensional global climate models. In J. E. Hansen and T. Takahashi (eds), *Climate Processes and Climate Sensitivity*, Geophysical Monograph 29, Maurice Ewing Vol. 5, Amer. Geophys. Union, Washington DC, pp. 58–72.

Dickinson, R. E., Henderson-Sellers, A., Kennedy, P. J., and Wilson, M. F. (1986) Biosphere/atmosphere transfer scheme (BATS) for the NCAR community climate model, *NCAR Technical Note TN-275 + STR*, 69pp.

Eagleson, P. S. (ed.) (1982) *Land Surface Processes in Atmospheric General Circulation Models*, Cambridge University Press, Cambridge, 560pp.

Gordon, C. T., and Stern, W. F. (1982) A description of the GFDL global spectral model, *Mon. Wea. Rev.*, **110**, 625–644.

Hansen, J., Russell, G., Rind, D., Stone, P., Lacis, A., Lebedeff, S., Ruedy, R., and Travis, L. (1983) Efficient three dimensional global models for climate studies: Models I and II, *Mon. Wea. Rev.*, **111**, 609–622.

Henderson-Sellers, A. (1987) Effects of change in land-use on climate in the humid tropics. In R. E. Dickinson (ed.), *The Geophysiology of Amazonia*, John Wiley & Sons, New York, pp. 463–493.

Manabe, S., and Bryan, K. (1985) CO_2 induced changes in a coupled ocean–atmosphere model and its palaeoclimatic implications, *J. Geophys. Res.*, **90**, 1689–1707.

Manabe, S., Bryan, K., and Spelman, M. J. (1979) A global ocean–atmosphere climate model with seasonal variation for future studies of climate sensitivity, *Dyn. Atmos. Oceans*, **3**, 393–426.

Matthews, E. (1983) Global vegetation and land use: New high-resolution data bases for climate studies, *J. Clim. Appl. Meteor.*, **22**, 474–487.

Mintz, Y. (1984) The sensitivity of numerically simulated climates to land-surface boundary conditions. In J. T. Houghton (ed.), *The Global Climate*, Cambridge University Press, Cambridge, pp. 79–106.

Ramanathan, V. (1981) The role of ocean–atmosphere interactions in the CO_2 climate problem, *J. Atmos. Sci.*, **38**, 918–930.

Rossow, W. B., Mosher, F., Kinsella, E., Arking, A., Desbois, M., Harrison, E., Minnis, P., Ruprecht, E., Sèze, G., and Summer, C. (1985) ISCCP cloud algorithm intercomparison, *J. Clim. Appl. Meteor.*, **24**, 877–903.

Sellers, P. J., Mintz, Y., Sud, Y. C., and Dalcher, A. (1986) A Simple Biosphere model (SiB) for use with general circulation models, *J. Atmos. Sci.*, **43**, 505–531.

Shukla, J., and Mintz, Y. (1982) Influence of land-surface evapotranspiration on the earth's climate, *Science*, **215**, 1498–1501.

Simmons, A. J., and Bengtsson, L. (1987) Atmospheric general circulation models: their design and use for climate studies. In M. E. Schlesinger (ed.), *Physically Based Climate Models and Climate Modelling*, Proceedings of a NATO ASI, Reidel, Dordrecht.

Washington, W. M., and Parkinson, C.L. (1986) *An Introduction to Three-Dimensional Climate Modelling*, University Science Books, Mill Valley, Calif., 422pp.

Washington, W. M., Semtner, A. J. Jr., Knight D. J., and Mayer, T. A. (1980) A general circulation model experiment with a coupled ocean, atmosphere and sea-ice model, *J. Phys. Oceanog.*, **10**, 1887–1908.

Wells, N. C. (1979) A coupled ocean–atmosphere experiment: the ocean response, *Quart. J. Roy. Meteor. Soc.*, **105**, 355–370.

Wilson, M. F., and Henderson-Sellers, A. (1985) Land cover and soils data sets for use in general circulation climate models, *J. Climatol.*, **5**, 119–143.

Appendix A Glossary

Absorption bands. Molecules absorb radiation by being excited into vibration and rotation. In the case of water vapour these absorption bands lie in the same region as the longwave radiation emitted by the Earth and are therefore of significance in climate studies. Carbon dioxide does not possess rotation bands, but its main vibration band also lies in this region as do the absorption bands of some trace gases such as chlorofluorocarbons and methane.

Absorptivity. The fraction of the radiation incident upon a body which the body absorbs is called its absorptivity; also known as absorptance, although this usually refers to a particular wavelength.

Advection. Horizontal transport, usually of energy, mass, etc., in the atmosphere or ocean.

Aerosol. A suspension of small solid particles or liquid droplets in the atmosphere. Aerosols can be both natural (e.g. volcanic) and man-made (e.g. industrial) in nature, and may have significant radiative effects on climate.

Albedo. From the Latin *albus*, meaning white. It is the reflected fraction of incident radiation. It is synonymous with hemispheric reflectance.

Aliasing. In the sampling of a continuous function to produce values at discrete points, the sampling frequency must be high enough to resolve the highest frequency present in the continuous function. If not, the high-frequency information will appear as a false enhancement of a related lower frequency. The high frequency signal is then said to be be aliased into a lower frequency.

Baroclinic. The atmosphere is baroclinic when isotherms are not parallel to isobars, i.e. there is a temperature gradient along the isobars.

Baroclinic waves. A disturbance in a smooth zonal flow which is equivalent barotropic will generate baroclinic waves. They are characteristic of the upper-level atmospheric flow in mid-latitudes.

Barotropic. The atmosphere is barotropic when there are horizontally uniform temperatures at all heights. Thus pressure gradients can exist but horizontal temperature gradients cannot. In an equivalent barotropic atmosphere horizontal temperature gradients may exist but isotherms must be parallel to the isobars.

Bottom water. The cold dense water which lies at the bottom of the oceans in contact with the ocean floor. It is formed in high latitudes as a result of the formation of sea-ice and has distinctly different characteristics from the water immediately above it.

Boundary layer. The lowest 1 km or so of the atmosphere. It is the location of all interactions between the Earth's surface (land, ocean and ice) and the free atmosphere.
Bowen ratio. The ratio of sensible heat flux to latent heat flux from a surface.
Clausius–Clapeyron (equation). The dependence of the saturation vapour content of the atmosphere upon its temperature.
Climate system. Variously defined. The atmosphere, hydrosphere, cryosphere, land surface and biosphere, i.e. all the components of the Earth's environment are, to a greater or lesser extent, components of the climate system.
Convection. When a parcel of air is warmer than its environment it will move upwards and carry energy with it. This is the process of convection.
Convergence. See **Divergence**.
Coriolis force. The apparent force experienced by an entity moving over a rotating body such as the Earth. On the surface of the Earth the direction of travel is deflected towards the right in the northern hemisphere and to the left in the southern hemisphere.
Cryosphere. A collective term describing the frozen water masses of the Earth. These are ice sheets, snow cover, glaciers, permafrost and sea-ice.
Degree day. Summer warmth can be expressed as the sum of degree days: the product excess of daily temperatures above a preset threshold which is often the threshold for plant growth (about 5° C) and the number of days exceeding the threshold by that amount. Winter cold can also be summarized this way except that the temperature depression below a threshold is used.
Degrees of freedom. Each observation in a random sample of size n can be compared with $n-1$ other observations and hence there are $n-1$ degrees of freedom.
Dependent variables. If a quantity y is a function of a quantity x then y is the dependent variable. It is dependent upon x. It follows that x is the independent variable.
Diffuse radiation. When radiation passes through the atmosphere, it is scattered in interactions with molecules and particles in the air. This scattering alters the direction of travel of the radiation progressively and randomly. After a great deal of scattering then the radiation becomes totally diffuse (i.e. the same quantity of radiation is travelling in all directions). Generally the radiation received at the Earth's surface is composed of a diffuse component and a direct (unscattered) component. The factor 1.66 is often introduced into calculations to allow for the larger distance (on average) travelled by diffuse radiation through an atmospheric layer. Thermal radiation is emitted from all directions within the atmosphere and therefore is totally diffuse to a good approximation.
Diffusion coefficient. This determines the rate of diffusion of a quantity along a gradient of that quantity.
Divergence. The 'spreading out' of a flow or the flux of a quantity away from a point in more than one direction. Convergence is its opposite.
Eddy fluxes. The flux of a quantity by means of turbulent disturbances propagating in a fluid. Such a flux is made possible by the process called eddy diffusion.
Effective temperature. The temperature of the Earth, derived by simple energy balance considerations assuming the Earth to be a black body emitter.
Emissivity. The ratio of the emittance from a body to that of a black body (perfect) emitter at the same temperature is the emissivity of that body.

Enstrophy. The root mean square of the vorticity of a body (usually a fluid).
Entrainment. The process by which fluid external to a parcel is incorporated into the parcel.
Equilibration time. The time taken for a component of the climate system to reach equilibrium with one or more of the other components.
Feedback. The phenomenon whereby the output of a system is fed into the input and the output is subsequently affected.
Flux. The flow of a quantity through a surface. The flux of, for example, energy is always *from* somewhere *to* somewhere else, i.e. its vector nature is important.
Fourier transform. A continuous function can be represented mathematically either by its value at many grid points or as the sum of many waves of differing frequency, amplitude and phase. The two representations are formally equivalent, i.e. they contain the same information. (A good analogy is a number and its logarithm to any base.) The Fourier transform of a variable is the frequency domain equivalent of its space domain representation. A fast Fourier transform (FFT) is a computer algorithm for moving between the frequency and spatial domains.
Gain. The gain of a system is a measure of the amplification of the input to the system. In its simplest form it equals output divided by input.
Gaussian quadrature. A particular type of numerical computation of a one-dimensional integral from the knowledge of individual values of it.
General circulation of the atmosphere. Two factors control the general circulation of the atmosphere: the energy imbalance between the equator (net absorption of energy) and the poles (net emission of energy) and the rotation of the Earth. In low latitudes the direct Hadley cell circulations transfer energy polewards but in the middle latitudes the rotation of the system causes a wave-like flow in the troposphere (Figure A.1). These Rossby waves travel in a westerly direction and energy transfer is via horizontal eddies which form the familiar depression systems and anticyclones of mid-latitude weather. In polar regions there are weak direct cellular flows but the seasonal variation in insolation from polar day to night dominates the pattern.

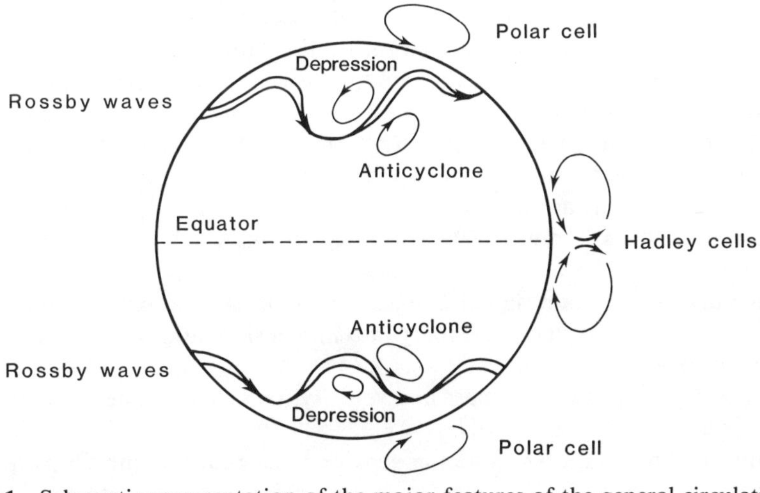

Fig A.1 Schematic representation of the major features of the general circulation of the atmosphere.

Gravity waves. Analogous to water waves. They occur in the atmosphere due to forcing by orography and manifest themselves as undulations in potential temperature surfaces.
Greenhouse effect. The effect of the atmosphere in re-radiating longwave radiation back to the surface of the Earth. It has nothing to do with glasshouses, which trap warm air at the surface.
Hadley cell. The thermally driven circulation comprising an upward motion at the ITCZ and downward motion in the subtropics. The air moves poleward at high altitude and equatorward near the surface.
Heat capacity. The energy required to change the temperature of a body by 1 K.
Hydrostatic equation. The relationship between pressure (p) and height (z) in the atmosphere. $\Delta p = -\varrho g \Delta z$.
Ideal gas law. The statement that the ratio of the product of pressure and volume of a gas to its temperature is a constant.
Infinite. Quite a lot—really an awful lot.
Independent variable. See **Dependent variable**.
Intertropical convergence zone (ITCZ). The zone of convergence between the two Hadley cells. Characterized by large amounts of vertical ascent.
Isothermal. Having the same temperature everywhere. Often used to describe the situation where no vertical temperature gradient exists in the atmosphere.
Isotropic. Having the same properties in all directions.
Lapse rate. The rate of decrease of temperature with height in the atmosphere.
Latent heat. The energy used in conversion between different phases of water. It is the energy used in evaporation of water (vaporization) or in the melting of ice (fusion).
Leaf area index (LAI). The area of foliage per unit area of ground. Usually only the upper side of the leaves is considered but an alternative definition includes both upper and lower leaf surfaces.
Little Ice Age. A period between the fifteenth and eighteenth centuries when temperatures were lower than today over many areas of the northern hemisphere. The timing differs from region to region and recently there has been some doubt cast upon whether a 'climatic epoch' really occurred.
Meridional circulation. The circulation of the atmosphere is dominated by meridional motions caused by the equator-to-pole temperature gradient.
Milankovitch mechanism. The orbital parameters of the Earth are constantly changing due to the influence of other planets. These changes in the orbital geometry of the Earth result in changes in the pattern of insolation at the Earth. This may provide a forcing agent for climate variations, the Milankovitch mechanism.
Mixing ratio (of water vapour). The quantity of water vapour (in kg) per kilogram of air.
Model noise. In climate models, inaccuracies in calculations, rounding and truncation errors introduce variability into model simulations which are classed as model noise.
Momentum. The product of mass and velocity. In a closed system, momentum is conserved.
Non-linear. An equation with dependent variable y containing terms in y^2 or

higher is non-linear. Such equations exhibit a response to perturbations which is not related in a constant fashion to the forcing.

Optical thickness. Radiation passing through a body is attenuated by a factor $e^{-\delta}$, where δ is the optical thickness of the body. It is a strong function of wavelength. Alternatively, the length of time required by any student of climate modelling to 'see the light'.

Oxygen isotope data. Because of their different masses water molecules containing different oxygen isotopes are evaporated and precipitated at different rates, The ratio of ^{18}O to ^{16}O can be used to indicate the nature of palaeoclimates since it is related to ocean temperatures.

Ozone hole. A description of the appearance, over the poles (most pronounced over Antarctica), in ~1985–86 of an area of distinctly reduced ozone levels. The cause of these reduced levels is unknown and may either be natural or man-induced. Confirmation of the mechanism will require further observation of ozone levels.

Parameterization. The method of incorporating a process by representation as a simplified function of some other fully resolved variables without explicitly considering the details of the process.

Planck function. The description of the amount of radiation emitted by a body at a given temperature as a function of the wavelength of the radiation.

Potential evaporation. Potential evaporation can be defined as the evaporation which would occur given free and plentiful supply of water. Clearly such a situation would occur only over water bodies and wet vegetation.

Potential temperature. In the atmosphere the potential temperature of a parcel of air is the temperature it would have if moved under adiabatic conditions (without gaining or losing energy) to a reference level (usually 1000 hPa). Potential temperature is more useful than actual temperature because the compressibility of gases means that temperature increases as pressure increases and vice versa.

Prognostic variable. A variable in a model which is utilized in prediction of itself at an advanced time point.

Richardson number. A measure of the stability of a fluid layer: a ratio of buoyancy to inertial forces.

Scattering. In the atmosphere the redirection of light waves due to reflection and diffraction by atmospheric molecules (Rayleigh) and cloud droplets (Mie).

Sensitivity. The sensitivity of a model to a perturbation. Usually described as a unit of response per unit change.

Solar constant. The amount of radiation from the Sun incident on a surface at the top of the atmosphere perpendicular to the direction of the Sun. Currently taken to be $1370\ W\ m^{-2}$ and known to be variable! Note that S can denote both $1370\ W\ m^{-2}$, one quarter of this and the instantaneous top-of-the-atmosphere solar flux at a particular location. Context usually indicates which is meant but beware confusing algebra.

Solar elevation. The angular distance above the horizon of the Sun at any time.

Solar zenith angle. The angular distance between the solar position and the true zenith (the point directly overhead). It is the complement of the elevation angle.

Spherical harmonic. For a variable defined on the surface of a sphere the natural method of representation in the frequency domain is by means of functions called spherical harmonics. These can be thought of as the extension to spherical coordinates of the concept of Fourier transformation (q.v.).

Stability. A measure of the capacity of a system to resist perturbation. The ability to recover the original position after displacement.

Stability (of the atmosphere). Consider the situation in Figure A.2(a) where the dashed line shows the critical lapse rate, γ_c. The dotted lines show both stable and unstable temperature profiles. If a parcel of air is displaced from the surface it will rise such that its temperature decreases at a rate given by γ_c. If the environmental lapse rate is stable, then the displaced parcel will be colder (and thus denser) than the surrounding air (Point A) and will sink back. In the unstable case the parcel will be warmer (less dense) than its surroundings (Point B) and will continue to rise.

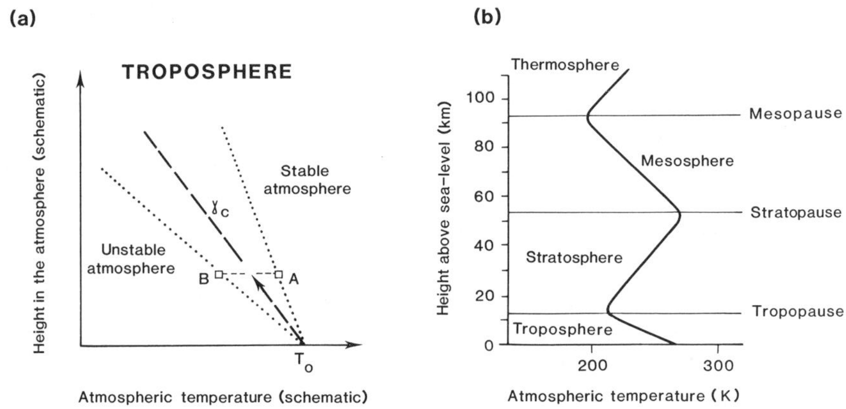

Figure A.2 (a) Stable and unstable tropospheric lapse rates as compared with a predetermined critical lapse rate (γ_c). (b) The four "regions" of the atmosphere.

Stefan–Boltzmann constant, σ, having a value $5.67 \times 10^{-8}\,\mathrm{W\,m^{-2}\,K^{-4}}$, the constant of proportionality in Stefan's law.

Stefan's law. This is the relationship between the amount of energy radiated by a body and its absolute temperature and is given by $E = \sigma T^4$ where E is in $\mathrm{W\,m^{-2}}$ and σ is the Stefan–Boltzmann constant.

Stratosphere. The zone of the atmosphere above the troposphere (Figure A.2(b)). Atmospheric temperature increases with height due to absorption of radiation by ozone.

Stream function. A description of the amount of volume transport taking place at a particular point (in space) in an ocean or atmospheric model per unit time. It has units of $\mathrm{m^3\,s^{-1}}$. In the ocean the term Sverdrup is often used where 1 Sverdrup $= 10^6\,\mathrm{m^3\,s^{-1}}$.

Thermodynamics. The science of the movement of heat. Usually concerned with calculations involving fluids.

Timestep. The base unit of temporal resolution in a numerical model.

Transitivity. The phenomenon whereby a system (in this case the climate) evolves from an initial state to another, different, state and stays there.

Transmissivity. The portion of the radiation, incident upon a body, which passes through it.
Troposphere. Lowest region of the atmosphere (Figure A.2(b)). The atmosphere is well mixed and temperature decreases with height. Clouds and weather systems are confined below its upper boundary, the tropopause, at a global average height of about 10 km.
Truncation. When a function, which is currently represented as a summation of many terms, is reduced in length and complexity by removing a number of small terms.
Variance. A measure of the spread of a set of results (about a mean). Used in climate modelling as a measure of the variability of the model climate.
Vorticity. The vorticity of a body is twice its angular velocity about the vertical. The Coriolis parameter $f = 2\Omega \sin \phi$ is the vorticity of the Earth. The absolute vorticity is the vorticity of the body plus the vorticity of the Earth at latitude ϕ. The potential vorticity of an air column is the absolute vorticity divided by the height of the column. Potential vorticity is always conserved.
Weak Sun–warm early Earth paradox. It is known that the solar constant was 20–30 per cent lower in the Earth's early history. Despite this, the surface temperature has never been as low as would be expected from simple calculations. It is possible that the larger amounts of CO_2 in the early atmosphere and different cloud and surface characteristics may have compensated for the lower solar luminosity.

Appendix B

Climate Models: Examples of Simple Microcomputer Software

All the programs described are available on IBM PC discs. Details from the authors.

This part of the text is 'learned' rather than 'taught'. There is a great deal of effort required to understand the levels of complexity included in the wide range of climate models. In order to help newcomers to climate modelling as much as possible, four simple models have been written and developed for microcomputers.

POSSIBLE PROJECTS

We do not intend to identify specific tasks. Answering many, if not all, of the proposed questions will aid understanding.

Beware, many are very hard!

The models

These are *very* much simpler models than many of the large numerical models described in the book. If you list them from the micro you will find that they only take up two or three pages, whereas a (fairly) simple GCM would produce a paper listing about 10 cm thick. They're also very simple to run—deceptively so. Your goal is to *understand* the literature upon which these models are based and to pose for yourselves questions about the validity and applicability of each model type.

1. *Daisyworld*. This is the parable world which illustrates biological homoeostasis. It is the simplest of the models and without doubt the hardest to understand.
2. *Anthropogenerated albedo change*. This model uses a fairly simple set of three equations, but is based upon a series of very difficult assessments of anthropogenerated land-use change and subsequent albedo and climate change. Its use should prompt you to read a wide range of literature.
3. *An EBM*. This is the EBM you saw in Chapter 3; now is the time for you to understand how it works. More importantly, you should try to understand what the inherent simplifications are in this, and all other, EBMs.

4. *CO_2 feedback using a simple ocean model*. This is the most complex of the four models. Its use should prompt you to read the literature about (a) the pre-industrial level of atmospheric carbon dioxide and its subsequent change; (b) climate model doubling CO_2 experiments; and (c) oceans and ocean models.

It is important and useful to compare and contrast aspects of all these different models. For example:

(a) Why is it that Daisyworld exhibits tolerance to surface albedo changes, while anthropogenerated land-use change seems to cause global temperature changes?

(b) Why does the EBM react so extremely to small changes in solar luminosity while Daisyworld maintains stability over a very wide range of solar luminosity changes?

(c) Do you understand the difference between an equilibrium climate change and a transient one? Which do each of these models compute? Why are they different? Which, if any, are wrong and why?

(d) The CO_2 sensitivity model is the only example model described here which contains an ocean. In fact the CO_2 model is almost completely dependent upon the ocean. If it is a true representation of a climate system, what is the point in all the others?

MODEL I DAISYWORLD: A SIMPLE BIOSPHERIC FEEDBACK CLIMATE MODEL

This model of the global environment is based on the paper by Watson and Lovelock (1983). This fictional world is a parable which its authors believe describes the fundamental characteristic of the biosphere—namely that it has a strong tendency to provide homoeostasis (internal stability).

It is based on an equation for the rate of plant growth, R, as a function of temperature, T (°C), which is derived from ecological observations, such that

$$R = 1 - 0.003265(22.5 - T)^2 \tag{B.1}$$

There are two types of plants on this fictional world: black daisies and white daisies. The comparative growth of these daisies is governed by a simple pair of differential equations which also form the basis of all 'prey–predator models' and are known as the Lotka–Volterra equations

$$\begin{aligned} \mathrm{d}a_w/\mathrm{d}t &= a_w(xR - D) \\ \mathrm{d}a_b/\mathrm{d}t &= a_b(xR - D) \end{aligned} \tag{B.2}$$

where a_w and a_b are the fractional areas covered by the white and black daisies respectively, x is the fractional area of fertile ground not covered by either species, R is the growth rate of daisies per unit time given by equation (B.1) and D is the death rate of daisies per unit time which can be chosen arbitrarily and in this version has a value of 0.3.

The model allows you to reproduce the behaviour of either a sterile world (without daisies) or worlds with one or both daisy species represented. The albedos of the daisies and the bare ground can be changed, and by changing the solar luminosity the curves given in Watson and Lovelock (1983) can be reproduced.

The very strong conclusion is that any world containing any living material which behaves in a way similar to these daisies is more likely to retain hospitable global temperatures than a planet without life. There are some questions which might be prompted by the parable of Daisyworld:

1. Is the parabolic growth rate of plants as a function of temperature (equation (B.1)) valid for all life?
2. Does life exhibit other controlling feedbacks, for example with atmospheric carbon dioxide?
3. Must all non-living feedbacks be 'disruptive' to the biological homoeostasis or are there powerful feedbacks in environments without life?

Bibliography

Allaby, M., and Lovelock, J. E. (1984) *The Greening of Mars*, André Deutsch, London, 165pp.

Lovelock, J. E. (1979) *Gaia. A New Look at Life on Earth*, Oxford University Press, Oxford, 157pp.

Lovelock J. E., and Watson, A. J. (1982) The regulation of carbon dioxide and climate. Gaia or geochemistry? *Planet. Space Sci.*, **30**, 793–802.

Owen, T., Cess, R. D., and Ramanathan, V. (1979) Enhanced CO_2 greenhouse to compensate for reduced solar luminosity on the early earth, *Nature*, **277**, 640–642.

Schneider, S. H. (1986) A goddess of the Earth?: the debate on the Gaia hypothesis—an editorial, *Climatic Change*, **8**, 1–4.

Walker, J. C. G., Hays, P., and Kasting, J. F. (1981) A negative feedback mechanism for the long-term stabilization of earth's surface temperature, *J. Geophys. Res.*, **86**, 9776–9782.

Watson, A. J., and Lovelock, J. E. (1983) Biological homoeostasis of the global environment: the parable of Daisyworld, *Tellus*, **35**, 285–288.

MODEL II MODELLING THE CLIMATIC IMPACT OF ANTHROPOGENERATED ALBEDO CHANGE

In this exercise a hypothesis proposed by Sagan *et al.* (1979) is examined using a simple climate model. The basis of the hypothesis is that, by changing the land surface, mankind is altering the surface albedo and hence (probably) the planetary albedo, and thus modifying the amount of energy available to the climate system so causing climatic change. Sagan *et al.* propose the use of a simple equation which states that the planetary albedo change from each land-use change, i, is given by

$$\Delta R_i = A_i I_i [F_i \Delta r_t + (1 - F_i)\Delta r_s] \tag{B.3}$$

where R is the planetary albedo, A is the area, I is the insolation factor, F is the cloud cover fraction, r_s is the surface albedo and Δ implies a change. The only term remaining is Δr_t which is the change in the planetary albedo in the presence of clouds. This is obtained using the equation

$$r_t = r_c + (1 - r_c)^2 r_s/(1 - r_s r_c) \tag{B.4}$$

where Sagan *et al.* assumed a cloud albedo, $r_c = 0.48$, in order to obtain the observed planetary albedo of 0.3 with a mean surface albedo of 0.1. Finally Sagan *et al.* assume that there is a linear relationship between the change in planetary albedo and the resultant change in surface temperature. From other model studies they use a scaled version of the relation which states if $\Delta R = 0.01$ then $\Delta T = 2$ K.

A number of stages must be gone through in order to be able to use and understand this simple model of climate change. These are:

1. Estimation of the current extent of each of the undisturbed land types.
2. Identification of all the current and likely anthropogenerated land-use changes.
3. Establishment of the associated albedo change (see e.g. Henderson-Sellers and Gornitz, 1984).
4. Assessment of the appropriateness of the linear relationship between planetary albedo and surface temperature.

The computer model will give you all the answers to these questions in your first run through the simulation, but complete understanding of the strengths and weaknesses of such a simple model requires you to think about each of the points listed above.

Bibliography

Dickinson, R. E. (1987) *The Geophysiology of Amazonia*, Wiley, New York, 526pp.

Hansen, J., Johnson, D., Lacis, A., Lebedeff, S., Lee, P., Rind, D., and Russell, G. (1981) Climate impact of increasing carbon dioxide, *Science*, **213**, 957–966.

Henderson-Sellers, A., and Gornitz, V. (1984) Possible climatic impacts of land cover transformations, with particular emphasis on tropical deforestation, *Climatic Change*, **6**, 231–257.

Sagan, C., Toon, O. B., and Pollack, J. B. (1979) Anthropogenic albedo changes and the Earth's climate, *Science*, **206**, 1363–1368.

MODEL III AN ENERGY BALANCE CLIMATE MODEL (EBM)

This type of climate model is a useful teaching/learning tool. The fundamentals of construction and use of EBMs were described in detail in Chapter 3. Another version of the model listed in Chapter 3 is considered here. Both hemispheres are now incorporated along with explicit representation of cloud and calculation of top-of-the-atmosphere fluxes. The EBM is formulated according to the equation

$$\text{(Solar radiation in)} = \text{(Longwave out)} + \text{(Transport out)} \qquad \text{(B.5)}$$

This condition is imposed for each latitude zone so that an equilibrium climate solution is maintained. Cloud amount is prescribed and is used in the calculation of the planetary albedo. The emitted longwave is recalculated from the final temperature distribution to produce the given distribution.

Program structure

The program is driven through a series of menus. The integration is initiated by choosing option 4 of the 'Main menu'. There is a series of other 'menus' which allow you to change different parts of the model.

Note that if you choose option 2 repeatedly in the final panel (after running the model to equilibrium) then you can 'customize' nearly all of the features of the model.

Aim of the EBM simulation exercise

The *main aim* of using this simple climate model is to acquaint you with two separate but complementary areas of current climatological research:
(a) climate modelling;
(b) satellite monitoring of climate.

Table B.1 Microcomputer EBM equilibrium climate

Latitude (°)	Surface temperature (°C)	Planetary albedo	Specified cloud amount	Emitted IR flux ($W\,m^{-2}$)	Absorbed solar flux ($W\,m^{-2}$)
85	−15.4	0.68	0.52	165	55
75	−14.9	0.68	0.58	165	58
65	−1.3	0.35	0.62	194	139
55	4.3	0.34	0.63	207	173
45	10.2	0.32	0.57	220	208
35	14.2	0.34	0.46	230	231
25	19.1	0.32	0.40	241	261
15	27.3	0.24	0.42	259	310
5	26.1	0.28	0.50	255	303
5	25.0	0.29	0.50	253	296
15	27.3	0.24	0.42	259	310
25	22.9	0.26	0.40	249	284
35	18.0	0.27	0.46	238	254
45	11.1	0.30	0.57	222	213
55	5.0	0.33	0.63	208	177
65	−8.1	0.54	0.62	180	99
75	−14.9	0.68	0.58	165	58
85	−15.4	0.68	0.52	165	55

A task which illustrates the increased complexity of the model is to try to discover the fundamental characteristics of the Earth's climate using as your source data satellite-observed fluxes at the top of the atmosphere.

1. Run the model with our suggested values and a solar constant fraction of 1 (i.e. the present day solar illumination). Note the temperature distribution and the emitted infrared and absorbed solar fluxes (see Table B.1).
2. Suppose a new satellite has observed the following zonally averaged fluxes.

Latitude zone	Emitted infrared flux ($\mathrm{W\,m^{-2}}$)	Absorbed solar flux ($\mathrm{W\,m^{-2}}$)
90–80°N	157	51
80–70°	159	55
70–60°	195	150
60–50°	209	187
50–40°	221	217
40–30°	225	227
30–20°	233	249
20–10°	251	295
10–0°	254	303
0–10°	251	296
10–20°	251	295
20–30°	240	169
30–40°	233	248
40–50°	223	223
50–60°	211	193
60–70°	162	64
70–80°	159	55
80–90°S	157	51

Can you force the climate model to simulate these fluxes by altering some or all of the features of the climate system? You have the freedom to change the following climate characteristics:

(a) *Longwave radiation formula*. You can change A or B or both in the equation:

$$L_{out} = A + BT + \text{(cloud term)} \tag{B.6}$$

(b) *Transport term*. You can change K in the energy transport formulation:

$$\text{Transport} = K(T - \bar{T}) \tag{B.7}$$

(c) *Cloud amounts*. By changing the zonally averaged cloud amount you can cool (with more cloud) or warm (less cloud) each zone. Also more cloud in a zone means less emitted infrared radiation because the cloud temperatures are lower than the surface temperatures.

Table B.2 Zonally averaged total mean monthly cloud amount (%) for 1979

	Month												
	Jan	Feb	Mar	Apr	May	June	July	Aug	Sept	Oct	Nov	Dec	Mean
Northern hemisphere													
90–85	89.7	92.8	89.2	63.5	29.9	32.1	31.3	32.1	50.0	64.2	82.8	83.8	61.8
85–80	65.9	65.8	63.9	48.8	25.1	30.4	31.3	33.0	43.6	53.6	66.1	61.5	49.1
80–75	50.8	50.8	48.9	42.3	30.2	37.5	42.8	45.8	49.7	50.5	58.0	48.2	46.3
75–70	50.3	49.7	44.6	45.1	43.9	49.2	50.3	59.0	60.2	53.3	53.7	47.2	50.5
70–65	48.5	44.5	44.8	48.5	55.7	63.2	64.0	71.0	68.2	58.4	55.5	48.2	55.9
65–60	53.7	50.8	55.3	55.4	60.9	65.1	66.7	71.6	69.9	67.1	61.6	57.3	61.3
60–55	59.7	58.6	60.9	59.2	62.0	64.0	65.5	67.7	69.0	72.1	70.9	63.1	64.4
55–50	59.9	62.7	60.2	61.8	60.3	63.3	65.2	63.4	64.7	66.3	68.7	62.2	63.2
50–45	60.4	64.2	59.5	62.1	58.2	58.5	57.9	56.4	54.2	62.2	65.4	64.4	60.3
45–40	61.1	65.4	60.2	64.4	56.9	56.0	52.9	50.4	47.1	58.0	59.3	64.2	58.0
40–35	62.5	65.7	61.2	62.0	55.9	54.1	47.3	42.2	44.1	50.0	54.7	59.2	54.9
35–30	56.1	57.4	56.1	53.0	52.1	48.5	43.6	38.2	39.1	39.4	47.3	48.6	48.3
30–25	49.5	46.7	49.8	46.3	48.7	46.4	45.6	40.9	36.8	33.2	41.8	41.8	43.9
25–20	42.1	38.4	40.8	38.9	44.4	50.5	50.2	48.0	40.8	32.4	40.5	37.4	42.0
20–15	38.6	32.8	34.3	37.4	46.1	58.9	58.7	56.5	48.2	40.1	43.0	37.7	44.4
15–10	39.1	35.4	37.8	44.9	55.3	71.0	73.5	68.0	61.0	54.8	53.2	42.8	53.1
10–5	52.5	52.4	54.3	64.0	71.7	83.6	83.0	72.1	69.3	68.5	70.1	58.3	66.6
5–0	64.0	66.3	66.2	68.4	69.7	75.7	67.7	58.1	52.4	59.4	63.5	63.4	64.6
Southern hemisphere													
0–5	67.4	69.1	67.0	64.3	57.8	63.2	58.1	61.6	62.8	68.7	69.5	66.7	64.7
5–10	69.8	68.9	67.2	56.9	52.3	54.8	52.5	54.6	57.7	62.0	67.0	68.6	61.0
10–15	69.7	67.0	67.7	51.1	48.2	48.9	48.3	49.7	51.6	56.0	65.8	64.5	57.4
15–20	65.8	64.7	62.1	46.6	46.5	43.7	43.0	42.1	42.5	47.8	54.7	57.1	51.4
20–25	58.5	58.2	55.6	50.3	48.1	46.3	41.3	41.4	43.2	46.4	48.6	51.1	49.1
25–30	54.2	56.3	54.7	54.6	52.2	52.1	47.0	48.1	49.2	50.6	51.8	50.7	51.8
30–35	54.0	58.8	58.4	58.8	58.9	54.8	54.3	57.0	58.1	57.5	56.1	53.8	56.7
35–40	57.7	64.7	67.2	64.8	62.3	59.6	59.8	64.7	64.8	66.0	63.3	58.8	62.8
40–45	58.6	67.0	67.9	63.4	62.7	62.8	64.0	69.0	69.9	69.3	66.8	59.2	65.1
45–50	61.6	65.5	65.7	63.7	61.0	63.3	68.3	72.1	73.5	71.7	68.9	61.1	66.4
50–55	66.6	66.6	66.5	63.7	62.5	64.8	69.1	72.6	75.9	73.4	72.2	63.3	68.1
55–60	69.8	69.7	70.2	67.0	66.3	71.2	70.3	71.9	72.9	69.5	68.4	63.5	69.2
60–65	65.6	72.4	72.6	71.0	67.1	73.8	68.7	62.1	58.2	53.8	51.9	52.9	64.2
65–70	48.9	64.4	69.1	70.3	58.8	74.0	74.6	62.5	59.4	46.5	40.3	35.9	58.7
70–75	40.8	60.0	68.2	73.5	67.8	85.2	84.5	79.7	72.2	55.5	41.7	30.6	63.3
75–80	34.1	65.6	76.7	82.7	78.4	83.4	84.2	85.3	79.0	63.5	37.8	22.8	66.1
80–85	24.7	61.5	82.7	87.9	86.4	84.8	83.4	87.2	81.4	69.1	34.4	16.5	66.7
85–90	28.4	76.8	94.1	96.0	97.0	89.7	89.6	93.7	91.1	86.9	51.4	32.2	77.2

(d) *Surface and cloud albedos*. Both of these can be changed. By increasing albedos you will reduce absorbed solar radiation and cool the climate. What values give you the best simulation?

Other possible simulation modes

1. Return to the basic model and impose a different *observed* cloud distribution. You could try the observed mean cloud amounts for 1979 given in Table B.2.

How do these values change (i) surface climate, (ii) satellite-observed fluxes? Can you modify climate features (as in the first exercise) to regain the original satellite fluxes?

2. Suppose CO_2 increases caused a decrease in cloud amount by between 10 and 20 per cent. How is the climate affected? Could satellite measurements of zonal fluxes be a way of detecting such climate changes?
3. The traditional EBM simulations were examinations of the sensitivity of the model to changes in the solar constant. This will give you a 'feel' for the basic characteristics of the model and how it responds to a simple perturbation. Find the percentage change in solar constant to initiate a complete glaciation of the Earth.

Other questions which will be prompted by these exercises and the associated references are:

1. If Sellers' (1969) rather than Budyko's (1969) formulation had been used for the transport term in equation (B.7), how would the results of the model change?
2. In what way have researchers suggested that the constants A and B in the longwave term should be modified to account for clouds (see Exercise 4 in Section 3.4)? How might the clouds affect the albedo in the shortwave term?
3. How does the 'atmospheric circulation' encapsulated in an EBM compare with that simulated by a GCM?
4. How might the long timescale effects of oceans be included in EBMs?

Bibliography

Budyko, M. I. (1969) The effect of solar radiation variations on the climate of the earth, *Tellus*, **21**, 611–619.

Cess, R. D. (1976) Climatic change, a reappraisal of atmospheric feedback mechanisms employing zonal climatology, *J. Atmos. Sci.*, **33**, 1831–1843.

North, G. R., Cahalan, R. F., and Coakley, J.A. (1981) Energy balance climate models, *Rev. Geophys. Space Phys.*, **19**, 91–121.

Sellers, W. D. (1969) A global climatic model based on the energy balance of the earth–atmosphere system, *J. Appl. Meteor.*, **8**, 392–400.

Warren, S. G., and Schneider, S. H. (1979) Seasonal simulation as a test for uncertainties in the parameterization of a Budyko–Sellers zonal climate model, *J. Atmos. Sci.*, **36**, 1377–1391.

MODEL IV CARBON DIOXIDE FEEDBACK USING A SIMPLE OCEAN MODEL

This model illustrates three important features of climate sensitivity and also illustrates a very wide range of other characteristics of both models and the real world.

The model is based on a simple box model of the global ocean. This ocean model consists of an upper mixed layer with a depth of 70 m coupled to a diffusive deep ocean. The exercise consists of giving the program the information it requires to perform a transient CO_2 climate sensitivity experiment. In this experiment carbon dioxide is added to the atmosphere at an exponential rate determined by the carbon dioxide value in parts per million by volume (ppmv) in 1850—which you provide—and two known values of 315 ppmv in 1958 and 338 ppmv in 1980. Over this 130-year period, the oceanic upper mixed layer temperature is computed by the model. To do this the equation

$$C_m \, \mathrm{d}\Delta T/\mathrm{d}t = \Delta Q - \lambda \, \Delta T - \Delta F \qquad \text{(B.8)}$$

must be solved for the change in mixed layer temperature, ΔT, as a function of time. The temperature changes because of the change in atmospheric CO_2 levels which change the heat flux by a value ΔQ. Heat loss, ΔF, occurs from the mixed layer into the deep, diffusive ocean and all other physical processes are assimilated into the feedback parameter, λ, which was defined in Section 1.4.4. The value of this is related to the values of ΔQ and ΔT after an instantaneous doubling of CO_2 (or 'switch-on') experiment when the system has reached equilibrium:

$$\Delta Q_{2x} = \lambda \, \Delta T_{2x} \qquad \text{(B.9)}$$

For example, 'switch-on' experiments give $\Delta Q_{2x} = 4.3$ W m^{-2}, which for values of ΔT_{2x} in the range of 1–5 K, gives values of λ in the range 0.8–4.0 W m^{-2} K^{-1}.

This first model computation follows the work of Wigley and Schlesinger (1985) and results in an evaluation of the temperature change (ΔT) of the ocean upper mixed layer. This is an interesting and important result but, as Hansen *et al.* (1985) point out, it is not the whole story.

At the end of any climatic perturbation, whether it be instantaneous or transient, the climate system requires some further time to regain equilibrium. This final change to the equilibrium temperature, ΔT_{eq} (defined as the difference between the resultant equilibrium temperature and an initial equilibrium temperature), differs from the calculated ΔT by an amount which is dependent upon the climatic feedback factor, f (as discussed in Chapter 1). This feedback factor is defined as the ratio of the equilibrium temperature with feedbacks to the equilibrium temperature without feedbacks. It is *exceedingly* difficult to understand, much less to model, this factor as it contains features such as clouds, the cryosphere, the biosphere, etc.

A number of assessments have been made of the climatic feedback factor, f, which is currently believed to lie between 1.2 and 3.6. (Note that an atmosphereless black body planet would have a feedback factor, f, of 1—cf. Figure 1.9.) In this exercise, this feedback factor, f, has been determined implicitly by the formulation used in equation (B.8) in terms of λ

$$f = 3.583/\lambda \qquad \text{(B.10)}$$

The result of any system having a feedback factor, f, greater than 1 is that its final equilibrium temperature will be greater than that calculated by a model which excludes these feedbacks.

The ocean box model in the exercise parameterizes all implicit climatic feedbacks in the $\lambda\ \Delta T$ term. At the end of the 130-year simulation, the temperature of the mixed layer has not yet attained its equilibrium value. The program offers you two values of the 130-year surface air temperature increment after this transient CO_2 increase experiment. The first air temperature increase is taken to be identical to the ocean mixed layer temperature increase. An alternative air temperature increase is calculated using the equation of Hansen *et al.* (1985).

$$T_0(X) = \ln(1.4\times10^{-6}X^3+0.005\ X^2+1.2X+1) - \ln(1.4\times10^{-6}X_0^3+0.005X_0^2+1.2X_0+1) \qquad \text{(B.11)}$$

where X is the CO_2 concentration at the end of the 130-year period and X_0 the value at the beginning (1830). Finally the equilibrium temperature (using both estimates for ΔT_0) is calculated from

$$\Delta T_{eq} = f\ \Delta T_0 \qquad \text{(B.12)}$$

This exercise prompts a myriad of questions pertaining to climate modelling and climate system dynamics:

1. What is the difference between a transient and an instantaneous (or switch-on) climate perturbation?
2. Are climate feedbacks calculable independently of an experiment or observation which retrieves ΔT_0 and ΔT_{eq}?
3. One of the major forcing factors in the model is ΔQ for doubling CO_2. This is derived from model results. Is this valid? Do they all agree?
4. The rate of carbon dioxide increase in the atmosphere appears to be crucial. Could we improve methods of estimating the pre-industrial atmospheric loading of CO_2?
5. There are important feedbacks entirely neglected by this model. In particular as ocean temperatures increase, the CO_2 which has been disssolved in the ocean is released back into the atmosphere, thus further increasing atmospheric CO_2 levels. How could this be parameterized and how important a term would it be?

Bibliography

Hansen, J., Lacis, A., Rind, D., Russell, G., Stone, P., Fung, I., Ruedy, R., and Lerner, J. (1984) Climate sensitivity: analysis of feedback mechanisms. In J. E. Hansen and K. Takahashi (eds), *Climate Processes and Climate Sensitivity*, Geophysical Monographs 29, Maurice Ewing Vol. 5, American Geophysical Union, 130–163; (especially Appendix).

Hansen, J., Russell, G., Lacis, A., Fung, I., Rind, D., and Stone, P. (1985) Climate response times: dependence on climate sensitivity and ocean mixing, *Science*, **229**, 857–859.

Harvey, L. D. D., and Schneider, S. H. (1985) Transient climate response to external forcing on 10^0–10^4 year time scales 2. Sensitivity experiments with a seasonal, hemispherically averaged, coupled atmosphere, land and ocean energy balance model, *J. Geophys. Res*, **90**, 2207–2222.

Luther, F. M., and Cess, R. D. (1985) Review of the recent carbon dioxide–climate controversy. In *The Potential Climatic Effects of Increasing Carbon Dioxide*, United States Department of Energy, Washington DC, pp. 321–335.

Oeschger, H., Siegenthaler, U., Schotterer, U., and Gugelmann, A. (1975) A box diffusion model to study the carbon dioxide exchange in nature, *Tellus*, **27**, 168–192.

Wigley, T. M. L. (1983) The pre-industrial carbon dioxide level, *Climatic Change*, **5**, 315–320.

Wigley, T. M. L., and Schlesinger, M. E. (1985) Analytical solution for the effect of increasing CO_2 on global mean temperature, *Nature*, **315**, 649–652.

General Bibliography

Abbott, E. A. (1963) *Flatland: A Romance of Many Dimensions*, (5th edn), Barnes and Noble, New York, 108pp.

Adem, J. (1965) Experiments aiming at monthly and seasonal numerical weather prediction, *Mon. Wea. Rev.*, **93**, 495–503.

Adem, J. (1979) Low resolution thermodynamic grid models, *Dyn. Atmos. Ocean.*, **3**, 433–451.

Allaby, M., and Lovelock, J. E. (1984) *The Greening of Mars*, André Deutsch, London, 165pp.

Alexander, R. C., and Mobley, R. L. (1974) Monthly average sea surface temperature and ice pack limbs over a $1° \times 1°$ global grid, *Res. Report 4-1310-APPA*, Rand Corporation, Santa Monica, California, 22pp.

Anthes, R. (1977) A cumulus parameterization scheme utilizing a 1-D cloud model, *Mon. Wea. Rev.*, **105**, 270–286.

Barron, E. J., and Washington, W. M. (1984) The role of geographic variables in explaining paleoclimates: Results from Cretaceous climate model sensitivity studies, *J. Geophys. Res.*, **69**, 1267–1279.

Barry, R. G., Henderson-Sellers, A., and Shine, K. P. (1984) Climate sensitivity and the marginal cryosphere. In Hansen, J. E., and Takahashi, T. (eds), *Climate Processes and Climate Sensitivity*, Maurice Ewing Series 5, American Geophysical Union, Washington, D.C., 221–237.

Berger, A. (1979) Insolation signature of quaternary climate changes, *Il Nuovo Cimento*, **2C(1)**, 63–87.

Berner, R. A., Lasaga, A. C., and Garrels, R. M. (1983) The carbonate-silicate geochemical cycle and its effect on atmospheric carbon dioxide over the last 100 million years, *Am. J. Sci.*, **283**, 641–683.

Boer, G. J., McFarlane, N. A., Laprise, R., Henderson, J. D., and Blanchet, J-P. (1984) The Canadian Climate Centre spectral atmospheric general circulation model, *Atmos-Ocean*, **22**, 397–429.

Bolin, B., Döös, B., Jäger, J., and Warrick, R. A. (1986) *SCOPE 29. The Greenhouse Effect, Climatic Change, and Ecosystems*, John Wiley and Sons, Chichester, 541pp.

Bourke, W., McAveney, B., Puri, K., and Thurling, R. (1977) Global modelling of atmospheric flow by spectral methods. In J. Chang (ed.), *Methods in Computational Physics*, **17** Academic Press, New York, 267–324.

Borisenkov, Ye. P., Tsvetkov, A. V., and Agaponov, S. V. (1983) On some characteristics of insolation changes in the past and the future. *Climatic Change*, **5**, 237–244.

Bowman, K. P. (1982) Sensitivity of an annual mean diffusive energy balance model with an ice sheet. *J. Geophys. Res.*, **87**, 9667–9674.

Bretherton, F. P. (1982) Ocean climate modeling. *Progress in Oceanography*, **II**, 93–129.

Broecker, W. S., and Van Donk, J. (1970) Insolation changes, ice volumes and the O^{18} record in deep sea cores, *Rev. Geophys.*, **8**, 169–196.

Broecker, W. S., Takahashi, T., Simpson, H. S., and Peng, T.-H. (1979) Fate of fossil fuel carbon dioxide and the global carbon budget. *Science*, **206**, 409–422.

Broecker, W. S., and Takahashi, T. (1984) Is there a tie between atmospheric CO_2 content and ocean circulation?, in Hansen, J. E., and Takahashi, T. (eds) *Climate Processes and Climate Sensitivity*, Maurice Ewing Series 5. American Geophysical Union, Washington, D.C., 314–326.

Bryan, K., Komro, F. G., Manabe, S., and Spelman, M. J. (1982) Transient climate response to increasing atmospheric carbon dioxide, *Science*, **215**, 56–58.

Bryson, R. A., and Goodman, B. M. (1980) Volcanic activity and climatic change. *Science*, **207**, 1041–1044.

Budyko, M. I. (1969) The effect of solar radiation variations on the climate of the Earth, *Tellus*, **21**, 611–619.

Carbon Dioxide Assessment Committee (1983) *Changing Climate*, Washington, D.C., National Academy Press, 490pp.

Cess, R. D. (1976) Climatic change: An appraisal of atmospheric feedback mechanisms employing zonal climatology, *J. Atmos. Sci.*, **33**, 1831–1843.

Cess, R. D. (1983) Arctic aerosols: Model estimates of interactive influences upon the surface–atmosphere clear-sky budget. *Atmos. Environ.*, **17**, 2555–2564.

Cess, R. S., and Goldenberg, S. D. (1981) The effect of ocean heat capacity upon global warming due to increasing atmospheric carbon dioxide, *J. Geophys. Res.*, **86**, 498–502.

Cess, R. D., Briegleb, B. P., and Lian, M. S. (1982) Low-latitude cloudiness and climate feedback: Comparative estimates from satellite data, *J. Atmos. Sci.*, **39**, 53–59.

Chang, J. (ed.) (1977) General circulation models of the atmosphere, *Methods in Computational Physics*, **17**, Academic Press, New York, 337pp.

Charlson, R. J., Lovelock, J. E., Andreae, M. O., and Warren, S. G. (1987) Ocean phytoplankton, atmospheric sulphur, cloud albedo and climate, *Nature*, **326**, 655–661.

Charney, J. G. (1975) Dynamics of deserts and drought in the Sahel, *Quart. J. Roy. Meteor. Soc.*, **101**, 193–202.

Chervin, R. M. (1980) On the simulation of climate and climate change with general circulation models, *J. Atmos. Sci.*, **37**, 1903–1913.

Chervin, R. M. (1981) On the comparison of observed and GCM simulated climate ensembles, *J. Atmos. Sci.*, **38**, 885–901.

Chervin, R. M., and Schneider, S. H. (1976) On determining the statistical significance of climate experiments with general circulation models, *J. Atmos. Sci.*, **33**, 405–412.

Chou, M.-D., Peng, L., and Arking, A. (1982) Climate studies with a multi-layer energy balance model, Part II: The role of feedback mechanisms in the CO_2 problem. *J. Atmos. Sci.*, **39**, 2657–2666.

Chylek, P., and Kiehl, J. T. (1981) Sensitivities of radiative–convective models, *J. Atmos. Sci.*, **38**, 1105–1110.

Climate Research Board (1979) *Carbon Dioxide and Climate: Scientific Assessment*, Washington, D.C., National Academy of Sciences, 22pp.

Cogley, J. G., and Henderson-Sellers, A. (1984) The origin and earliest state of the Earth's hydrosphere, *Rev. Geophys. Space. Phys.*, **22**, 131–175.

Deardorff, J. (1978) Efficient prediction of ground temperature and moisture with inclusion of a layer of vegetation, *J. Geophys. Res.*, **83**, 1889–1903.

DeBruin, H. A. R. (1983) A model for the Priestley-Taylor parameter α, *J. Clim. Appl. Meteorol.*, **22**, 572–578.

Dickinson, R. E. (1981) Convergence rate and stability of ocean–atmosphere coupling schemes with a zero-dimensional climate model, *J. Atmos. Sci.*, **38**, 2112–2120.

Dickinson, R. E. (1982) Modeling climate changes due to carbon dioxide increases, in Clark, W. C. (ed.) *Carbon Dioxide Review 1982*, New York, Oxford University Press, 101–133.

Dickinson, R. E. (1983) Land surface processes and climate–surface albedos and energy balance, *Adv. Geophys.*, **25**, 305–353.

Dickinson, R. E. (1984) Modeling evapotranspiration for three dimensional global climate models in Hansen, J. E. and Takahashi, T. (eds), *Climate Processes and Climate Sensitivity*, Maurice Ewing Series 5, American Geophysical Union, Washington, D.C., 58–72.

Dickinson, R. E. (1985) Climate sensitivity. In Manabe, S. (ed.) *Issues in Atmospheric and Oceanic Modelling, Part A, Climate Dynamics*, Advances in Geophysics, Volume 28, Academic Press, New York, pp. 99–129.

Dickinson, R. E. (ed.) (1987) *The Geophysiology of Amazonia*, John Wiley and Sons, New York, 526pp.

Dickinson, R. E., Henderson-Sellers, A., Kennedy, P. J., and Wilson, M. F. (1986) Biosphere–atmosphere transfer scheme (BATS) for the NCAR community climate model, *NCAR Technical Note TN-275+STR*, 69pp.

Douglas, R. G., and Woodruff, F. (1981) Deep sea foraminifera. In C. Emilian (ed.), *The Sea*, vol. 7, Wiley Interscience, New York, pp. 1233–1327.

Eagleson, P. S. (ed.) (1982) *Land Surface Processes in Atmospheric General Circulation Models*, Cambridge University Press, Cambridge, 560pp.

Eddy, J. A., Gilliland, R. L., and Hoyt, D. V. (1982) Changes in the solar constant and climatic effects, *Nature*, **300**, 689–693.

Fels, S. B., Mahlman, J. D., Schwarzkopf, M. D., and Sinclair, R. W. (1980) Stratospheric sensitivity to perturbations in ozone and carbon dioxide: Radiative and dynamical response, *J. Atmos. Sci.*, **37**, 2265–2297.

Foukal, P. V. (1980) Solar luminosity variation in directly observable time scales: observational evidence and basic mechanisms, in *Sun and Climate*, CNES/CNRS/DGRST Conference Proceedings, 275–284.

GARP (1975) The physical basis of climate and climate modelling, *GARP Publication Series No. 16*, WMO/ICSU, Geneva.

Gates, W. L. (1979) The effect of the ocean on the atmospheric general circulation, *Dyn. Atmos. Ocean.*, **3**, 95–109.

Gates, W. L. (1985) Modelling as a means of studying the climate system. In MacCracken, M. C. and Luther, F. M. (eds) *Projecting the climatic effects of increasing carbon dioxide*, DOE/ER-0237, US Dept. of Energy, Washington, D.C., pp. 57–79.

Gentilli, J. (ed.) (1971) Climates of Australia and New Zealand, *World Survey of Climatology*, Elsevier, Amsterdam, 405pp.

Gilchrist, A. (1983) Increased carbon dioxide concentrations and climate: The equilibrium response, in Bach, W. *et al.* (eds) *Carbon Dioxide: Current Views and Developments in Energy/Climate Research*, D. Reidel.

Gilliland, R. L. (1980) Solar luminosity variations, *Nature*, **286**, 838.

Gilliland, R. L. (1982) Modeling solar variability, *Astrophys. J.*, **253**, 399–405.

Golitsyn, G. S., and Mokhov, I. I. (1978) Sensitivity estimates and the role of clouds in simple models of climate, *Izvestiya, Atmospheric and Ocean Physics*, **14**, 569–576.

Gordon, C. T., and Stern, W. F. (1982) A description of the GFDL global spectral model, *Mon. Wea. Rev.*, **110**, 625–644.

Green, J. S. A. (1970) Transfer properties of the large scale eddies and the general circulation of the atmosphere, *Quart. J. Roy. Meteor. Soc.*, **96**, 157–185.

Hammer, C. U. (1977) Past volcanism revealed by Greenland Ice Sheet impurities, *Nature*, **270**, 482–486.

Hammer, C. U., Clausen, H. B., and Dansgaard, W., (1980) Greenland ice sheet evidence of post-glacial volcanism and its climatic impact, *Nature*, **288**, 230–235.

Hansen, J. E., Wang, W.-C., and Lacis, A. A. (1978) Mount Agung eruption provides test of a global climate perturbation, *Science*, **199**, 1065–1068.

Hansen, J. E., Lacis, A., Lee, P., and Wang, W.-C. (1980) Climatic effect of atmospheric aerosols, *Annals of the New York Academy of Sciences*, **338**, 575–586.

Hansen, J., Johnson, D., Lacis, A., Lebedeff, S., Lee, P., Rind, D., and Russell, G. (1981) Climate impact of increasing carbon dioxide, *Science*, **213**, 957–966.

Hansen, J., Russell, G., Rind, D., Stone, P., Lacis, A., Lebedeff, S., Ruedy, R., and Travis, L. (1983) Efficient three-dimensional global models for climate studies: Models I and II, *Mon. Wea. Rev.*, **111**, 609–622.

Hansen, J., Lacis, A., Rind, D., Russell, G., Stone, P., Fung, I., Ruedy, R., and Lerner, J. (1984) Climate sensitivity: analysis of feedback mechanisms in Hansen, J. E., and Takahashi, T. (eds), *Climate Processes and Climate Sensitivity*, Maurice Ewing Series 5, American Geophysical Union, Washington, D.C., 368pp.

Hansen, J., Russell, G., Lacis, A., Fung, I., Rind, D., and Stone, P. (1985) Climate response times: dependence on climate sensitivity and ocean mixing, *Science*, **229**, 857–859.

Hare, K. F. (1983) Climate and Desertification: A Revised Analysis, *World Climate Programme*, **44**, 149pp.

Hartmann, D. L. (1984) On the role of global-scale waves in ice–albedo and vegetation–albedo feedback, in Hansen, J. E., and Takahashi, T. (eds) *Climate Processes and Climate Sensitivity*, Maurice Ewing Series 5, American Geophysical Union, Washington, D.C., 18–28.

Hartmann, D. L., and Short, D. A. (1979) On the role of zonal asymmetries in climate change, *J. Atmos. Sci.*, **36**, 519–528.

Harvey, L. D. D., and Schneider, S. H. (1985) Transient climate response to external forcing on 10^0–10^4 year time scales 2. Sensitivity experiments with a seasonal, hemispherically averaged, coupled atmosphere, land, and ocean energy balance model, *J. Geophys. Res.*, **90**, 2207–2222.

Hasselmann, K. (1976) Stochastic climate models, Part 1. Theory, *Tellus*, **28**, 473–485.

Hasselmann, K. (1979) On the signal-to-noise problem in atmospheric response studies, in *Meteorology over Tropical Oceans*, Royal Meteorological Society, 251–259.

Hayashi, Y. (1982) Confidence intervals of a climatic signal, *J. Atmos. Sci.*, **39**, 1895–1905.

Held, I. M. (1982) Climate models and the astronomical theory of the ice ages, *Icarus*, **50**, 449–461.

Held, I. M. and Suarez, M. J. (1974) Simple albedo feedback models of the icecaps, *Tellus*, **26**, 613–629.

Held, I. M., and Suarez, M. J. (1978) A two level primitive equation model designed for climate sensitivity experiments, *J. Atmos. Sci.*, **35**, 206–229.

Henderson-Sellers, A. (1987) Effects of change in land-use on climate in the humid tropics. In R. E. Dickinson (ed.) *The Geophysiology of Amazonia*, John Wiley and Sons, New York, pp. 463–493.

Henderson-Sellers, A., and Gornitz, V. (1984) Possible climatic impacts of land cover transformations, with particular emphasis on tropical deforestation, *Climatic Change*, **6**, 231–256.

Henderson-Sellers, A., and Robinson, P. J. (1986) *Contemporary Climatology*, Longman, Harlow, 439pp.

Henderson-Sellers, A., and Wilson, M. F. (1983) Surface albedo for climate modelling, *Rev. Geophys. Space Phys.*, **21**, 1743–1778.

Hibler, W. D., III (1984) The role of sea ice dynamics in modeling CO_2 increases, in Hansen, J. E., and Takahashi, T. (eds) *Climate Processes and Climate Sensitivity*, Maurice Ewing Series 5, American Geophysical Union, Washington, D. C., 238–253.

Hibler, W. D., and Bryan, K. (1984) Ocean circulation: Its effects on seasonal sea-ice simulations, *Science*, **224**, 489–491.

Hirschboeck, K. K. (1980) A new worldwide chronology of volcanic eruptions, *Palaeogeography, Palaeoclimatology, Palaeoecology*, **29**, 223–241.

Hoffert, M. I., Flannery, B. P., Callegari, A. J., Hsieh, C. T., and Wiscombe, W. (1983) Evaporation-limited tropical temperatures as a constraint on climate sensitivity, *J. Atmos. Sci.*, **40**, 1659–1668.

Houghton, J. T. (ed.) (1984) *The Global Climate*, Cambridge University Press, Cambridge, 233pp.

Hummel, J. R., and Kuhn, W. B. (1981) An atmospheric radiative-convective model with interactive water vapour transport and cloud development, *Tellus*, **33**, 372–381.

Hunt, B. G. (1973) Zonally symmetric global General Circulation Models with and without the hydrologic cycle, *Tellus*, **25**, 337–354.

Imbrie, J., and Imbrie, K. P. (1979) *Ice Ages: solving the mystery*, Macmillan, London, 224pp.

Jäger, J. (1983) *Climate and Energy Systems. A Review of their Interactions*, Wiley, 231pp.

Jaeger, L. (1976) Monatskarten des Niederschlags für die ganze Erde, *Bericht Deutsche Wetterdienst*, **18**, no. 139.

Katz, R. W. (1982) Statistical evaluation of climate experiments with general circulation models: A parametric time series modeling approach, *J. Atmos. Sci.*, **39**, 1446–1455.

Katz, R. W. (1983) Statistical procedures for making inferences about precipitation changes simulated by an atmospheric general circulation model, *J. Atmos. Sci.*, **40**, 2193–2201.

Keepin, W., Mintzer, I., and Kristofersan, L. (1986) Emission of CO_2 into the atmosphere. The rate of release of CO_2 as a function of future energy developments. In Bolin, B., Döös, B., Jäger, J., and Warrick, R. A. (eds) *SCOPE 29. The Greenhouse Effect, Climatic Change, and Ecosystems*, John Wiley and Sons, Chichester, pp. 35–92.

Kelly, P. M., and Sear, C. B. (1984) The climatic impact of explosive volcanic eruptions, *Nature*, **311**, 740–743.

Kiehl, J. T., and Ramanathan, V. (1983) CO_2 radiative parameterization used in climate models: Comparison with narrow band models and with laboratory data, *J. Geophys. Res.*, **88**, 5191–5202.

Kutzbach, J. E., and Otto-Bliesner, B. L. (1982) The sensitivity of the African-Asian monsoonal climate to orbital parameter changes for 9000 years B.P. in a low-resolution general circulation model, *J. Atmos. Sci.*, **39**, 1177–1188.

Labitzke, K., Naujokat, B., and McCormack, M.P. (1983) Temperature effects on the stratosphere of the April 4 1982 eruption of El Chichon, Mexico, *Geophys. Res. Lett.*, **10**, 24–27.

Lamb, H. H. (1970) Volcanic dust in the atmosphere; with a chronology and assessment of its meteorological significance, *Phil. Trans. R. Soc. (London) Series A.*, **266**, 425–533 (see also the updates given by Lamb in *Clim. Monit.*, **6**, 57–67, and *Clim. Monit.*, **12**, 76–90, 1983.).

Leith, C. E. (1973) The standard error of time-average estimates of climatic means, *J. Appl. Meteor.*, **12**, 1066–1069.

Lian, M. S., and Cess, R. D. (1977) Energy balance climate models: A reappraisal of ice-albedo feedback, *J. Atmos. Sci.*, **34**, 1058–1062.

Lindzen, R. S., Hou, A. Y., and Farrell, B. F. (1982) The role of convective model choice in calculating the climate impact of doubling CO_2, *J. Atmos. Sci.*, **39**, 1189–1205.

Livezey, R. E., and Chen, W. Y. (1983) Statistical field significance and its determination by Monte Carlo techniques, *Mon. Weather Rev.*, **111**, 46–59.

Lovelock, J. E. (1979) *Gaia. A new look at life on Earth*, Oxford University Press, Oxford, 157pp.

Lovelock, J. E., and Watson, A. J. (1982) The regulation of carbon dioxide and climate. Gaia or geochemistry? *Planet. Space Sci.*, **30**, 793–802.

Luther, F. M., and Cess, R. D. (1985) Review of the recent carbon dioxide-climate controversy. In *The Potential Climatic Effects of Increasing Carbon Dioxide*, United States Department of Energy, Washington, D. C., pp. 321–335.

Lynch, P. (1987) Techniques of initialization, *Weather*, **42**, 66–70.

MacCracken, M. C., and Ghan, S. (1987) *Design and use of zonally-averaged climate models*, UCRL-94338, University of California, Livermore CA, 44pp.

MacCracken, M. C, Ellis, J. S., Ellsaesser, H. W., Luther, F. M., and Potter, G. L. (1981) *The Livermore Statistical Dynamical Climate Model*, Lawrence Livermore National Laboratory, (UCID-19060).

Madden, R. A., and Shea, D. J. (1978) Estimates of the natural variability of time-averaged temperatures over the United States. *Mon. Wea. Rev.*, **106**, 1695–1703.

Manabe, S., and Möller, F. (1961) On the radiative equilibrium and heat balance of the atmosphere, *Mon. Wea. Rev.*, **89**, 503–532.

Manabe, S., and Strickler, R. F. (1964) Thermal equilibrium of the atmosphere with a convective adjustment, *J. Atmos. Sci.*, **21**, 361–385.

Manabe, S., and Bryan, K. (1969) Climate calculations with a combined ocean atmosphere model, *J. Atmos. Sci.*, **26**, 786–789.

Manabe, S., Bryan, K., and Spelman, M. J. (1979) A global ocean–atmosphere climate model with seasonal variation for future studies of climate sensitivity, *Dyn. Atmos. Oceans*, **3**, 393–426.

Manabe, S., and Bryan, K. (1985) CO_2 induced changes in a coupled ocean–atmosphere model and its palaeoclimatic implications, *J. Geophys. Res.*, **90**, 1689–1707.

Manabe, S. (ed.) (1985) *Issues in Atmospheric and Oceanic Modelling, Part A., Climate Dynamics*, Advances in Geophysics, Volume 28, Academic Press, New York, 591pp.

Manabe, S., and Stouffer, R. J. (1980) Sensitivity of a global climate model to an increase of CO_2 concentration in the atmosphere, *J. Geophys. Res.*, **85**, 5529–5554.

Manabe, S., and Wetherald, R. T. (1967) Thermal equilibrium of the atmosphere with a given distribution of relative humidity, *J. Atmos. Sci.*, **24**, 241–259.

Manabe, S., and Wetherald, R. T. (1975) The effect of doubling the CO_2 concentration on the climate of a general circulation model, *J. Atmos. Sci.*, **32**, 3–15.

Manabe, S., and Wetherald, R. T. (1980) On the distribution of climate change resulting from an increase in CO_2-content of the atmosphere, *J. Atmos. Sci.*, **37**, 99–118.

Manabe, S., Wetherald, R. T., and Stouffer, R. J. (1981) Summer dryness due to an increase of atmospheric CO_2 concentration, *Climatic Change*, **3**, 347–386.

Manabe, S., and Wetherald, R. T. (1982) Simulation of global cloud cover. In *Cloud/ Radiation Interaction* ICSU/WMO Meeting in Dublin 1982 WCP-34, WMO, Geneva, pp. 1–18.

Matthews, E. (1983) Global vegetation and land use: New high-resolution data bases for climate studies, *J. Clim. Appl. Meteor.*, **22**, 474–487.

McClatchey, R. A., Fenn, R. W., Selby, J. E. A., Volz, F. E., and Garing, J. S. (1972) *Optical properties of the atmosphere* (3rd edn), Environmental Research Papers No. 411, Air Force Cambridge Research Laboratories.

McNaughton, K. G., and Jarvis, P. G. (1983) Predicting effects of vegetation changes on transpiration and evaporation, in *Water Deficits and Plant Growth*, Vol. **VII**, New York, Academic Press, 1–47.

Michael, P., Hoffert, M., Tobias, M., and Tichler, J. (1981) Transient climate response to changing carbon dioxide concentration, *Climatic Change*, **3**, 137–153.

Milne, A. A. (1928) *The House at Pooh Corner*, Methuen, London, 176pp.

Mintz, Y. (1984) The sensitivity of numerically simulated climates to land-surface boundary conditions. In Houghton, J. T. (ed.), *The Global Climate*, Cambridge University Press, Cambridge, 79–106.

Mitchell, J. F. B. (1983) The seasonal response of a general circulation model to changes in CO_2 and sea temperatures *Q. J. R. Meteorol. Soc.*, **109**, 113–152.

Mitchell, J. F. B., and Lupton, G. (1984) A $4 \times CO_2$ integration with prescribed changes in sea surface temperatures, *Progress in Biometeorology*, **3**, 353–374.

Mitchell, J. M., Jr. (1983) Empirical modeling of effects of solar variability, volcanic events and carbon dioxide on global-scale average temperature since AD 1880, in McCormac, B. M. (ed.) *Weather and Climate Responses to Solar Variations*, Colorado Associated University Press, 265–273.

Mokhov, I. I. (1981) Effect of CO_2 on the thermal regime of the earth's climatic system, *Meteorologiya i Gidrologiya*, **4**, 24–34.

National Research Council (1982) *Carbon dioxide and climate: A second assessment. Report of the CO_2/Climate review panel*, National Academy Press, Washington, D.C., 72 pp.

Newell, R. E., and Deepak, A. (eds) (1982) Mount St. Helens Eruptions of 1980: atmospheric effects and potential climatic impact, *NASA* SP-458, NASA Scientific and Technical Information Branch, Washington, D.C., 119pp.

Newhall, C. G., and Self. S. (1982) The Volcanic Explosivity Index (VEI): an estimate of explosive magnitude for historical volcanism, *J. Geophys. Res.*, **87**, 1231–1238.

Newkirk, G., Jr. (1983) Variations in solar luminosity, *Ann. Rev. Astron. Astrophys.*, **21**, 429–467.

North, G. R. (1975) Theory of energy balance climate models, *J. Atmos. Sci.*, **32**, 3–15.

North, G. R., Cahalan, R. F., and Coakley, J. A. (1981) Energy-balance climate models, *Rev. Geophys. Space Phys.*, **19**, 91–122.

North, G. R., Mengel, J. G., and Short, D. A. (1983) Simple energy balance model resolving the seasons and the continents: application to the astronomical theory of the ice ages, *J. Geophys. Res.*, **88**, 6576–6586.

North, G. R., Mengel, J. G., and Short, D. A. (1984) On the transient response patterns of climate to time-dependent concentrations of atmospheric CO_2, in Hansen, J. E., and Takahashi, T. (eds) *Climate Processes and Climate Sensitivity*, Maurice Ewing Series 5, American Geophysical Union, Washington, D.C., 164–170.

Oerlemans, J. (1982) Glacial cycles and ice-sheet modeling, *Climatic Change*, **4**, 353–374.

Oerlemans, J., and van der Veen, C. J. (1984) *Ice sheets and climate*, Reidel, Dordrecht, 217pp.

Oerlemans, J., and Vernekar, A. D. (1981) A model study of the relation between Northern Hemispheric glaciation and precipitation, *Contrib. Atmos. Phys.*, **54**, 352–361.

Oeschger, H., Siegenthaler, U., Schotterer, U., and Gugelmann, A. (1975) A box diffusion model to study the carbon dioxide exchange in nature, *Tellus*, **27**, 168–192.

Ohring, G., and Adler, S. (1978) Some experiments with a zonally averaged climate model, *J. Atmos. Sci.*, **35**, 186–205.

Oliver, R. C. (1976) On the response of hemispheric mean temperature to stratospheric dust: an empirical approach, *J. Appl. Meteorol.*, **15**, 933–950.

Oort, A. H. (1983) *Global Atmospheric Circulation Statistics, 1958–1973*, NOAA Prof. Pap. 14, US Dept. of Commerce, Washington, D.C., 180pp + 24 fiche.

Oort, A. H., and Peixóto, J. P. (1983) Global angular momentum and energy balance requirements from observations, *Adv. Geophys.*, **25**, 355–490.

Owen, T., Cess, R. D., and Ramanathan, V. (1979) Enhanced CO_2 greenhouse to compensate for reduced solar luminosity on the early earth, *Nature*, **277**, 640–642.

Palmén, E. (1949) Meridional circulations and the transfer of angular momentum in the atmosphere, *J. Meteor.*, **6**, 429–430.

Parker, D. E. (1985) The influence of the Southern Oscillation and volcanic eruptions on temperature in the tropical troposphere, *J. Clim.*, **5**, 273–282.

Parker, D. E., and Brownscombe, J. L. (1983) Stratospheric warming following the El Chichon volcanic eruption, *Nature*, **301**, 406–408.

Parkinson, J. H., Morrison, L. V., and Stephenson, F. R. (1980) The constancy of the solar diameter over the past 250 years, *Nature*, **28**, 548–551.

Parry, M. L. (1978) *Climatic change, agriculture and settlement*, Dawson, Folkestone, 214pp.

Peixóto, J. P., and Oort, A. H. (1984) Physics of climate, *Rev. Mod. Phys.*, **56**, 365–430.

Pittock, A. B. (1978) A critical look at long-term sun–weather relationships, *Rev. Geophys. Space Phys.*, **16**, 400–420.

Pittock, A. B. (1983) Solar variability, weather and climate: an update, *Q. J. R. Meteorol. Soc.*, **109**, 23–57.

Pivovarova, Z. I. (1977) *Radiation Characteristics of Climate in the USSR*, Leningrad, 355 pages, Gidrometeoizdat (in Russian).
Platt, C. M. R., Reynolds, D. W., and Abshire, N. L. (1980) Satellite and lidar observations of the albedo, emittance and optical depth of cirrus compared to model calculations, *J. Atmos. Sci.*, **108**, 195–204.
Potter, G. L., and Cess, R. D. (1984) Background tropospheric aerosols: incorporation within a statistical dynamical climate model, *J. Geophys. Res.*, **89**, 9521–9526.
Potter, G. L., and Gates, W. L. (1984) A preliminary intercomparison of the seasonal response of two atmospheric climate models, *Mon. Wea. Rev.*, **112**, 909–917.
Potter, G. L., Ellsaesser, H. W., MacCracken, M. C., and Luther, F. M. (1979) Performance of the Lawrence Livermore Laboratory zonal atmospheric model, in *Proc. GARP JOC Study Conference on Climate Models, GARP Publication Series No. 22 (Vol. 2)*, WMO, Geneva, 995–1001.
Potter, G. L., Ellsaesser, H. W., MacCracken, M. C., and Mitchell, C. S. (1981) Climate change and cloud feedback: the possible radiative effects of latitudinal redistribution, *J. Atmos. Sci.*, **38**, 489–493.
Quiroz, R. S. (1983) The climate of the 'El Niño' winter of 1982–1983—a season of extraordinary climatic anomalies, *Mon. Wea. Rev.*, **111**, 1685–1706.
Ramanathan, V. (1977) Interactions between ice-albedo, lapse-rate, and cloud-top feedbacks: An analysis of the nonlinear response of a GCM climate model, *J. Atmos. Sci.*, **34**, 1885–1897.
Ramanathan, V. (1981) The role of ocean–atmosphere interactions in the CO_2 climate problem, *J. Atmos. Sci.*, **38**, 918–930.
Ramanathan, V., and Coakley, J. A., Jr. (1978) Climate modeling through radiative-convective models, *Rev. Geophys. Space Phys.*, **16**, 465–489.
Ramanathan, V., Singh, H. B., Cicerone, R. J., and Kiehl, J. T. (1985) Trace gas trends and their potential role on climate change, *J. Geophys. Res.*, **90**, D3, 5547–5566.
Rampino, M. R., and Self, S. (1984) Sulphur rich volcanic eruptions and stratospheric aerosols, *Nature*, **310**, 677–679.
Rind, D., and Lebedeff, S. (1984) *Potential Impacts of Increasing Atmospheric CO_2 with Emphasis on Water Availability and Hydrology in the United States*, Report Prepared for the Environmental Protection Agency, NASA Goddard Space Flight Center Institute for Space Studies, New York, NY, 96pp.
Robock, A. (1978) Internally and externally caused climate change, *J. Atmos. Sci.*, **35**, 1111–1122.
Robock, A. (1983a) Ice and snow feedbacks and the latitudinal and seasonal distribution of climate sensitivity, *J. Atmos. Sci.*, **40**, 986–997.
Robock, A. (1983b) The dust cloud of the century, *Nature*, **301**, 373–374.
Rossow, W. B., Henderson-Sellers, A., and Weinreich, S. K. (1982) Cloud-feedback—a stabilizing effect for the early Earth, *Science*, **217**, 1245–1247.
Rossow, W. B., Mosher, F., Kinsella, E., Arking, A., Desbois, M., Harrison, E., Minnis, P., Ruprecht, E., Sèze, G., and Summer, C. (1985) ISCCP cloud algorithm intercomparison, *J. Clim. Appl. Meteor.*, **24**, 877–903.
Rowntree, P., and Walker, J. (1978) The effects of doubling the CO_2 concentration radiative-convective equilibrium, in Williams, J. (ed.), *Carbon Dioxide, Climate and Society*, New York, Pergamon Press, 181–192.
Sagan, C., Toon, O. B., and Pollack, J. B. (1979) Anthropogenic albedo changes and the Earth's climate, *Science*, **206**, 1363–1368.
Saltzman, B. (1978) A survey of statistical dynamical models of terrestrial climate, *Advances in Geophysics*, **20**, 183–304.
Saltzman, B. (ed.) (1983) *Theory of Climate*, Advances in Geophysics Volume 25, Academic Press, New York, 505pp.
Schlesinger, M. E. (1983) A review of climate models and their simulation of CO_2-induced warming, *Int. J. Environ. Stud.*, **20**, 103–114.

Schlesinger, M. E. (ed.) (1987) *Physically Based Climate Models and Climate Modelling*, Proceedings of a NATO ASI, Reidel, Dordrecht.

Schlesinger, M. E., and Mitchell, J. F. B. (1985), Model projections of the equilibrium climatic response to increased carbon dioxide. In *Projecting the climatic effects of increasing carbon dioxide*, United States Department of Energy, DOE/ER 0237, Washington, D.C., pp. 81–148.

Schneider, S. H. (1972) Cloudiness as a global climatic feedback mechanism: the effects on the radiation balance and surface temperature of variations in cloudiness, *J. Atmos. Sci.*, **29**, 1413–1422.

Schneider, S. H. (1986) A goddess of the Earth?: the debate on the Gaia hypothesis—an editorial, *Climatic Change*, **8**, 1–4.

Schneider, S. H., and Dickinson, R. E. (1974) Climate modeling, *Rev. Geophys. Space Phys.*, **12**, 447–493.

Schneider, S. H., Washington, W. M., and Chervin, R. M. (1978) Cloudiness as a climatic feedback mechanism: Effects on cloud amounts of prescribed global and regional surface temperature changes in the NCAR GCM, *J. Atmos. Sci.*, **35**, 2207–2221.

Schneider, S. H., and Thompson, S. L. (1981) Atmospheric CO_2 and climate: Importance of the transient response, *J. Geophys. Res.*, **86**, 3135–3147.

Schutz, C., and Gates, W. L. (1971) Global climatic data for surface, 800 mb, 400 mb: January, Advanced Research Projects Agency, Rept. R-915-ARPA, Rand Corp., Santa Monica, 173pp.

Schutz, C., and Gates, W. L. (1972) Global climatic data for surface, 800 mb, 400 mb: July, Advanced Research Projects Agency, Rept. R-1029-ARPA, Rand Corp., Santa Monica, 180pp.

Sedlacek, W. A., Mroz, E. J., Lazrus, A. L., and Gandrud, B. W. (1983) A decade of stratospheric sulphate measurements compared with observations of volcanic eruptions, *J. Geophys. Res.*, **88**, 3741–3776.

Seidel, S., and Keyes, D. (1983) Can we delay a greenhouse warming?, in *The Effectiveness and Feasibility of Options to Slow a Build-up of Carbon Dioxide in the Atmosphere*, Strategic Studies Staff, Office of Policy Analysis, Office of Policy and Resources Management, EPA, Washington, D.C. 20460.

Self, S., Rampino, M. R., and Barbera, J. J. (1981) The possible effects of large 19th century volcanic eruptions on zonal and hemispheric surface temperatures, *J. Volcanology and Geothermal Res.*, **11**, 41–60.

Sellers, P. J., Mintz, Y., Sud, Y. C., and Dalcher, A. (1986) A Simple Biosphere model (SiB) for use with general circulation models, *J. Atmos. Sci.*, **43**, 505–531.

Sellers, W. D. (1969) A global climatic model based on the Energy balance of the Earth–atmosphere system, *J. Appl. Met.*, **8**, 392–400.

Sellers, W. D. (1973) A new global climate model, *J. Appl. Meteor.*, **12**, 241–254.

Sellers, W. D. (1976) A two-dimensional global climate model, *Mon. Wea. Rev.*, **194**, 233–248.

Semtner, A. J., Jr. (1984a) On modelling the seasonal thermodynamic cycle of sea ice in studies of climate change, *Climatic Change*, **6**, 27–38.

Semter, A. J., Jr. (1984b) Development of efficient, dynamical ocean–atmosphere models for climate studies, *J. Clim. Appl. Meteor.*, **23**, 353–374.

Shine, K. P., and Henderson-Sellers, A. (1983) Modelling climate and the nature of climate models: a review, *J. Climatol.*, **3**, 81–94.

Shukla, J., and Mintz, Y. (1982) Influence of land-surface evapotranspiration on the earth's climate, *Science*, **215**, 1498–1501.

Shuttleworth, W. J., and Calder, I. R. (1979) Has the Priestley-Taylor equation any relevance to forest evaporation? *J. Appl. Meteorol.*, **18**, 639–646.

Shutts, G. J. (1983) Parameterization of travelling weather systems in a simple model of large-scale atmospheric flow. In B. Saltzman (ed.), *Theory of Climate*, Advances in Geophysics, Vol. 25, Academic Press, New York, pp. 117–172.

Simkin, T., Sibert, L., McClelland, L., Bridge, D., Newhall, C., and Latter, J. H. (1981) *Volcanoes of the World*, Stroudsburg, Pennsylvania, Hutchinson Ross Publishing Company, 233pp.

Simmons, A. J., and Bengtsson, L. (1987) Atmospheric general circulation models: their design and use for climate studies. In Schlesinger, M. E. (ed.) *Physically Based Climate Models and Climate Modelling*, Proceedings of a NATO ASI, Reidel, Dordrecht.

Slingo, A. (ed.) (1985) *Handbook of the Meteorological Office 11-layer atmospheric general circulation model. Volume 1: Model description*, Dynamical Climatology Technical Note 29, U.K. Meteorological Office (Met O20), 155pp.

Smagorinsky, J. (1983) The beginnings of numerical weather prediction and general circulation modeling: early recollections. In B. Saltzmann (ed.) *Theory of Climate* Academic Press, New York, pp. 3–38.

Smith, E. A., von der Haar, T. H., Hicket, J. R., and Maschhoff, R. (1983) The nature of short-period fluctuations in solar irradiance received by the Earth, *Climatic Change*, **5**, 211–235.

Somerville, R. C. J., and Remer, L. A. (1984) Cloud optical thickness feedbacks in the CO_2 climate problem, *J. Geophys. Res.*, **89**, 9668–9672.

Sonett, C. P. (1984) Very long solar periods and the radiocarbon record, *Rev. Geophys. Space Phys.*, **22**, 239–254.

Spelman, M. J., and Manabe, S. (1984) Influence of oceanic heat transport upon the sensitivity of a model climate, *J. Geophys. Res.*, **89**, 571–586.

Stone, P. H. (1973) The effects of large scale eddies on climatic change, *J. Atmos. Sci.*, **30**, 521–529.

Stone, P. H., and Miller, D. A. (1980) Empirical relations between seasonal changes in meridional temperature gradients and meridional fluxes of heat, *J. Atmos. Sci.*, **37**, 1708–1721.

Stuiver, M. (1980) Solar variability and climatic change during the current millennium, *Nature*, **286**, 868–871.

Stuiver, M., and Quay, P. D. (1980) Changes in atmospheric carbon-14 attributed to a variable sun, *Science*, **207**, 11–19.

Taylor, K. E. (1980) The roles of mean meridional motions and large-scale eddies in zonally averaged circulations, *J. Atmos. Sci.*, **37**, 1–19.

Taylor, B. L., Gal-Chen, T., and Schneider, S. H. (1980) Volcanic eruptions and long-term temperature records: an empirical search for cause and effect, *Q. J. R. Meteorol. Soc.*, **106**, 175–199.

Thompson, S. L., and Schneider, S. H. (1979) A seasonal zonal energy balance climate model with an interactive lower layer, *J. Geophys. Res.*, **84**, 2401–2414.

Thompson, S. L., and Schneider, S. H. (1982) Carbon dioxide and climate: The importance of realistic geography in estimating the transient temperature response, *Science*, **217**, 1031–1033.

Trenberth, K. E. (1983) Interactions between orographically and thermally forced planetary waves, *J. Atmos. Sci.*, **40**, 1126–1153.

Turco, R. P., Toon, O. B., Ackerman, T., Pollack, J. B., and Sagan, C. (1983) Nuclear winter: Global consequences of multiple nuclear explosions, *Science*, **222**, 1283–1292.

US National Academy of Sciences (1975) *Understanding Climatic Change, a Program for Action*, Washington, D.C., 239pp.

Walker, J. C. G., Hays, P., and Kasting, J. F. (1981) A negative feedback mechanism for the long-term stabilization of earth's surface temperature, *J. Geophys. Res.*, **86**, 9776–9782.

Wang, W-C., and Stone, P. H. (1980) Effects of ice-albedo feedback on global sensitivity in a one-dimensional radiative–convective model, *J. Atmos. Sci.*, **37**, 545–552.

Warren, S. G., and Schneider, S. H. (1979) Seasonal simulation as a test for uncertainties in the parameterization of a Budyko–Sellers zonal climate model, *J. Atmos. Sci.*, **36**, 1377–1391.

Washington, W. M., and Meehl, G. A. (1983) General circulation model experiments on the climatic effects due to a doubling and quadrupling of carbon dioxide concentration, *J. Geophys. Res.*, **88**, 6600–6610.

Washington, W. M., and Meehl, G. A. (1984) Seasonal cycle experiment on the climate sensitivity due to a doubling of CO_2 with an atmospheric general circulation model coupled to a simple mixed layer ocean model, *J. Geophys. Res.*, **89**, 9475–9503.

Washington, W. M., and Parkinson, C. L. (1986) *An Introduction to Three-Dimensional Climate Modelling*, University Science Books, Mill Valley, Calif., 422pp.

Washington, W. M., Semtner, A. J., Jr., Meehl, G. A., Knight, D. J., and Meyer, T. A. (1980) A general circulation experiment with a coupled atmosphere, ocean and sea ice model, *J. Phys. Oceanogr.*, **10**, 1887–1908.

Watson, A. J., and Lovelock, J. E. (1983) Biological homeostasis of the global environment: the parable of Daisyworld, *Tellus*, **35**, 285–288.

Wells, N. C. (1979) A coupled ocean–atmosphere experiment: the ocean response, *Quart. J. Roy. Meteor. Soc.*, **105**, 355–370.

Wetherald, R. T., and Manabe, S. (1975) The effects of changing the solar constant on the climate of a general circulation model, *J. Atmos. Sci.*, **32**, 2044–2059.

Wetherald, R. T., and Manabe, S. (1980) Cloud cover and climate sensitivity, *J. Atmos. Sci.*, **37**, 1485–1510.

Wetherald, R. T., and Manabe, S. (1981) Influence of seasonal variation upon the sensitivity of a model climate, *J. Geophys. Res.*, **86**, 1194–1204.

Wigley, T. M. L. (1983) The pre-industrial carbon dioxide level, *Climatic Change*, **5**, 315–320.

Wigley, T. M. L., and Schlesinger, M. E. (1985) Analytical solution for the effect of increasing CO_2 on global mean temperature, *Nature*, **315**, 649–652.

Williams, L. D., Wigley, T. M. L., and Kelly, P. M. (1980) *Climatic trends at high northern latitudes during the last 4000 years compared with ^{14}C fluctuations, in Sun and Climate*, CNES/CNRS/DGRST Conference Proceedings, 11–20.

Williamson, D. L. (1983) Description of NCAR Community Climate Model (CCMOB), *NCAR Technical Note TN-210+STR*, 88pp.

Willson, R. C., Gulkis, S., Janssen, M., Hudson, H. S., and Chapman, G. C. (1981) Observations of solar irradiance variability, *Science*, **21**, 200–202.

Wilson, M. F., and Henderson-Sellers, A. (1985) Land cover and soils data sets for use in general circulation climate models, *J. Climatol.*, **5**, 119–143.

Wilson, M. F., Henderson-Sellers, A., Dickinson, R. E., and Kennedy, P. J. (1987) Sensitivity of the biosphere–atmosphere transfer scheme (BATS) to the inclusion of variable soil characteristics, *J. Clim. Appl. Meteor.*, **26**, 341–362.

Wiin-Nielsen, A. C., and Sela, J. (1971) On the transport of quasi-geostrophic potential vorticity, *Mon. Wea. Rev.*, **99**, 447–459.

Woods, J. D. (1984) The upper ocean and air–sea interactions, in Houghton, J. T. (ed.) *The Global Climate*, Cambridge University Press, Cambridge, 79–106.

World Climate Research Programme (1983) Report of the Experts Meeting on Aerosols and Their Climatic Effects, *World Climate Programme*, **55**, 101pp.

World Meteorological Organization (1982) WMO Global Ozone Research and Monitoring Project, Report No. 14 of the *Meeting of Experts on Potential Climatic Effects of Ozone and Other Minor Trace Gases*, Boulder, Colorado, 13–17 September 1982, Geneva, WMO.

World Meteorological Organization (WMO) (1986) *Report of the International Conference on the assessment of the role of carbon dioxide and of other greenhouse gases in climate variations and associated impacts*, Villach, Austria, 9–15 October, 1985, WMO No. 661, Geneva, Switzerland, 78pp.

Yeh, T.-C., Wetherald, R. T., and Manabe, S. (1983) A model study of the short-term climatic and hydrologic effects of sudden snow-cover removal, *Mon. Wea. Rev.*, **111**, 1013–1024.

Index

(Figure and Table references are in italics)